FRUIT AND
VEGETABLE PROCESSING
Organisations and Institutions

W0193036

FRUIT AND VEGETABLE PROCESSING
Organisations and Institutions

By

Suman Bhatti

Uma Varma

Associate Professor and Head,
Department of Extension Education,
IC. College of Home Science
CCS Haryana Agricultural University, Hisar

CBS

CBS Publishers & Distributors Pvt. Ltd.

New Delhi • Bengaluru • Chennai • Kochi • Kolkata • Mumbai
Hyderabad • Nagpur • Patna • Pune • Vijayawada

ISBN: 81-239-0404-5

First Edition: 1995
Reprint: 1997, 2000, 2006, 2007, 2012, 2017

Published by **Satish Kumar Jain** and produced by **Varun Jain** for
CBS Publishers & Distributors Pvt. Ltd.,
4819/XI Prahlad Street, 24 Ansari Road, Daryaganj, New Delhi - 110002
delhi@cbspd.com, cbspubs@airtelmail.in • www.cbspd.com
Ph.: 23289259, 23266861, 23266867 • Fax: 011-23243014

Corporate Office: 204 FIE, Industrial Area, Patparganj, Delhi - 110 092
Ph: 49344934 • Fax: 011-49344935
E-mail: publishing@cbspd.com • publicity@cbspd.com

Branches:
• *Bengaluru:* 2975, 17th Cross, K.R. Road, Bansankari 2nd Stage,
  Bengaluru - 70 • Ph: +91-80-26771678/79 • Fax: +91-80-26771680
  E-mail: cbsbng@gmail.com, bangalore@cbspd.com
• *Chennai:* No. 7, Subbaraya Street, Shenoy Nagar, Chennai - 600030
  Ph: +91-44-26681266, 26680620 • Fax: +91-44-42032115
  E-mail: chennai@cbspd.com
• *Kochi:* Ashana House, 39/1904, A.M. Thomas Road, Valanjambalam,
  Ernakulum, Kochi • Ph: +91-484-4059061-65
  Fax: +91-484-4059065 • E-mail: cochin@cbspd.com
• *Kolkata:* 6-B, Ground Floor, Rameshwar Shaw Road, Kolkata - 700014
  Ph: +91-33-22891126/7/8 • E-mail: kolkata@cbspd.com
• *Mumbai:* 83-C, Dr. E. Moses Road, Worli, Mumbai - 400018
  Ph: +91-9833017933, 022-24902340/41 • E-mail: mumbai@cbspd.com

Representatives:

• Hyderabad: 0-9885175004   • Nagpur: 0-9021734563
• Patna: 0-9334159340   • Pune: 0-9623451994
• Vijayawada: 0-9000660880

Printed at:
J.S. Offset Printers, Delhi (India)

FOREWORD

Of late, food processing has been engaging considerable attention of planners and policy makers because of its potential contribution to economic development of rural areas. Low-cost food processing technologies offer excellent opportunities for women in production of processed foods through promotion of cottage, small scale and big industries. Today only one per cent of total fruits and vegetables are processed in 3,000 food industries in the country. Fruit plantation and vegetable cultivation in extensive areas in the country has been triggered by green revolution. Both fruits and vegetables are perishable. Their seasonal glut in the market leads to colossal wastage. Therefore, it is in the fitness of things that fruit and vegetable processing is developed for economic benefit of growers and national prosperity.

The doctoral research study findings documented in this book "Fruit and Vegetable processing Organisations and Institutions" clearly sheds light on organisational structure, production, research and training; use of technology; linkages with concerned organisations and institutions; role of men and women for different jobs in food processing units of different sizes. As far as I know, the information presented in this book appears to be the first of its kind in the country. This information can be useful for bringing about desirable improvements while planning women's contribution in food processing at different levels, i.e. manual, skilled, managerial and marketing jobs.

It is clearly revealed that at different organisational levels women are employed less than men and that too for manual work only. Production, and that too of the pickle is the major role being performed by various organisations. Multiple products are produced by some big units only. Most of the organisations have been found using only one technology, i.e., use of salt. The interorganisational linkages for research and development, use of technology, financial inputs, training and dissemination of technology are negligible. Women's performance in managerial and marketing roles at organisational and institutional level is low as compared to men.

In the end, I congratulate Dr. (miss) Suman Bhatti and Dr. (mrs.) Uma Varma, the authors of this publication, for having initiatd a useful study in this new area of food processing industry. I am sure that further research work in this important field will be continued by the prospective scholars.

T. VERMA

Assistant Director General (Home Science)
Indian Council of Agricultural Research
Krishi Anusandhan Bhavan,
Pusa-Gate, New Delhi

and

Former Dean,
College of Home Science,
CCS Haryana Agricultural University, Hisar.

LIST OF TABLES

LIST OF ILLUSTRATIONS

LIST OF PLATES

CONTENTS

1

INTRODUCTION

The Green Revolution and subsequent efforts through the application of science and technology for increasing food production in India have brought self-reliance in food. The impetus given by the Government, State Agricultural Universities, State Departments of Agriculture and other organisations through the evolution and introduction of numerous hybrid varieties of cereals, legumes and vegetables and improved agricultural practices have resulted in increased food production. However, the nation still faces the problem of the use of improper methods for the storage of food stuffs, leading to great wastage of the food produced. Such losses in the food front aggravate the existing syndromes of under-nutrition and malnutrition.

Fruits and vegetables, which are among the perishable commodities, are important ingredients in the human dietaries. Due to their high nutritive value, they make significant nutritional contribution to human well-being. They are the cheaper and better source of the protective foods. If they can be supplied in fresh or preserved form throughout the year for human consumption, the national picture will improve greatly.

The perishable fruits and vegetables are available as seasonal surpluses during certain parts of the year in different regions and are wasted in large quantities due to absence of facilities and know-how for proper handling, distribution, marketing and storage. Furthermore, massive amounts of the perishable fruits and vegetables, produced during a particular season result in a glut in the market and become scarce during other seasons. Neither can they all be consumed in fresh condition nor sold at economically viable prices.

Therefore, food processing has been engaging the attention of planners and policy makers as it can contribute to the economic development of rural population. The utilization of resources both material and human is one of the ways of improving the economic status of family. As women constitute 50 per cent of population, they also play an important role in the economic upliftment of the family. Traditionally, women handle food and are familiar with the skills of food processing. In order to improve the status of women and rural food processing, low cost indigenous food processing technologies offer excellent opportunities for women in production of processed foods.

In the post-green revolution era, eventhough food grains have been taken care of, fruits and vegetables for want of simple technologies of processing, preservation and transport to various places of need, have suffered post-harvest losses, estimated to be nearly 35% only 1% of the total fruits and vegetables produced are processed in the 3,000 food industries in the country (Das, 1991). All forms of preserved fruits are in the reach of only the urban elite, and the rural masses who produce more than 90% of these fruits and vegetables are usually deprived of their usage.

India produces a number of fruits and vegetables, and their combined production is 80 million tonnes. About 35 per cent of the total production is unfortunately wasted due to inadequate facilities for processing (Anonymous, 1992). Despite such a large production, their processing is yet to be developed properly. The processing includes pre-processing of fruits and vegetables before these are fit to be used for final conversion into processed foods. Delay in the use of harvested food takes away its freshness, palatability, appeal and nutritive value. Tropical fruits are luscious, juicy and pulpy.

They cannot be plucked early, cold-stored or subjected to controlled and long drawn out process as is possible in the case of fruits grown in temperate or colder regions. They have to be harvested very near their optimum maturity and processed or consumed promptly as they ripen because they require special attention and techniques.

The food preservation and processing industry has now become more of a necessity than being a luxury. It has an important role in the conservation and better utilization of fruits and vegetables. In order to avoid the glut and utilise the surplus during the season, it is necessary to employ modern methods to extend storage life for better distribution and also processing techniques to preserve them for utilisation in the off season in both large scale and home scale.

Food preservation includes all procedures whereby a food material is either placed in such an environment or else processed in such a way as to increase its keeping quality for longer period of time. Various methods of fruit preservation serve as means for carrying fruits and vegetables not only from season to season, but also from places of abundant production to places of little or no production.

The rural homemakers who play a considerable role in food production have not been exposed to modern methods of preservation. Fruit and vegetable processing is being done at different levels. At home level by individual families, at small scale by small scale industry, Khadi and village industry and at large scale by big industry. Traditionally, women handle food and are familiar with the skills of food processing. Appropriate food processing technologies are being developed by different research and training institutes for transmission to the users. Low cost food processing technologies can offer excellent opportunities for women in production of processed foods in rural areas. Development of appropriate food processing industry in rural area depends on availability of suitable technology and institutions for imparting the training.

It is difficult to exactly quantify the role of rural women in fruit and vegetable processing. Inadequate or partial information about rightful access to development resources and services signifies the importance of having a fuller understanding of the role and contributions of women so that extension services may accordingly be attuned to fully integrate them in the development of food processing. The study of different linkages seems to be necessary to gather this requisite information. For this reason, the present study was proposed with the following objectives:-

1. To study the role performance and linkages of different organisations and institutions working for fruits and vegetable processing.
2. To identify the appropriate food processing technologies for dissemination among the rural families.
3. To study the crucial factors influencing food processing at different levels.
4. To suggest the strategy for promotion of participation of rural women in food processing.

SCOPE OF THE STUDY

The present study is an attempt to analyze the role analysis and linkages of fruit and vegetable processing organisations and institutions. Efforts have been made to focus role analysis and linkages of large scale, small scale, training, co-operative and Khadi gramudyog units, rural families and development of a strategy for promoting participation of women in food processing. Since no study has been conducted on such an aspect, therefore, this study will provide some important and basic information and role analysis and linkages of various organisations at different levels and institutions. The finding of this study will also be of some use to the policy makers engaged in the promotion of various programmes for development of women by removing the various constraints encountered by rural families particularly by women at different levels. This study will act as a torch bearer for organisation's personnel.

LIMITATIONS OF THE STUDY

The present study was undertaken as a student research programme and hence it had natural limitations of resources, particularly time and money. The problems become more acute when a field study is done by a female student, requiring lot of field visits. Due to this reason, the study was confined to three districts only with a small sample of two large scale, two small scale, two training units, one cooperative and one Khadi gramudyog units and 65 rural families.

Further, this study was also prone to the limitations of most of the social science researches which are based on the recale response of the subject. Hence, the objectivity of the study was limited to their ability to recall and also their honesty in furnishing required information.

ORGANIZATION

This dissertation is divided into six chapters. Following introductory portion in the first chapter, the second chapter deals with the theoretical orientation and conceptual model. Review of literature is being dealt in the third chapter. The methodology adopted for the selection of respondents, research locale, operationalisation and measurement of variables and statistical techniques have been described in the fourth chapter. The fifth chapter presents the findings and discussion. The summary of the study and its action implications have been given in the last chapter.

2

THEORETICAL ORIENTATION

The main purpose of this chapter is to systematically portray the framework. A comprehensive attempt has been made to review and present the literature with reference to the concept of fruit and vegetable processing as-

Chronological development of food processing in India.

Administration of fruit and vegetable processing at different levels.

Conceptual model of role performance and linkages of fruit and vegetable processing at different levels.

CHRONOLOGICAL DEVELOPMENT OF FOOD PROCESSING IN INDIA

Fruit and vegetable processing was first started in an organised manner in 1857, mainly to make pickles and chutneys with a view to meeting the export requirement and canning of fruits and vegetables was started in 1927. The introduction of modern techniques of processing and preservation by addition of chemical preservatives could also be said to have been started from the same time. During the period 1927-1940, the processing and preservation was mainly in the manufacture of soft drinks like squashes, juices, cordials, barley water etc. From 1940 onwards the industry diversified the product mix and it started making canned fruits and vegetables, jams, jellies, and marmalades, tomato products, fruit juices, etc.

In the initial stages of developement from 1927 onwards, the tendency was to locate the manufacturing units mainly in the consuming centres like Bombay, Calcutta, Madras, etc. Although this tendency, which is helpful for marketing finished products, still continues, the efforts made by various developmental agencies in the country have resulted in installation and commissioning of fruit and vegetable preservation units in the fruit growing areas in order to retain the maximum flavour and aroma of the raw materials. This is a commendable development and it is likely that in the years to come more and more preservation and processing units will be set up in the growing areas in the rural parts of the country, opening up new avenues for rural employment. As in most of the consuming city (centres), the manufacturers used to make purchase of raw materials from the markets of these centres, and as a result, direct link between the growers and processors was not established. Of late, this tendency has reversed and now the processors are trying to establish contacts with the growers directly so that they can get raw materials of adequate quality and the growers, in turn, can get a fair price for their produce. This has happened particularly with regard to pineapple, tomato, green pea and lime to the advantage of both growers and processors. In addition, there will be a gain also for the rural people resulting in all round rural development.

Khadi and village Industries commission (KVIC) is the nodal agency meant for rural industrialisation by promoting food based industries. KVIC implements its programme through state KVI boards, cooperative societies, registered institutions and individuals and extends needed assistance to programme implementing agencies.

ADMINISTRATION OF FRUIT AND VEGETABLE PROCESSING AT DIFFERENT LEVELS

Fruits and vegetables are processed at different levels by the governmental organisation and by non-governmental organisation. By the government processing is done at central, state and district levels. (Fig.2.1).

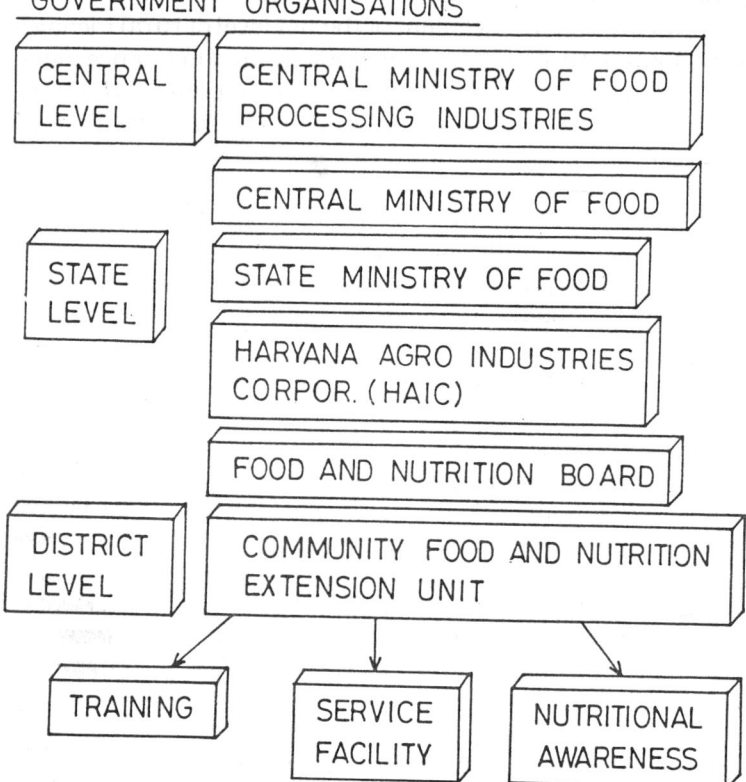

Fig. 2.1 : Administration of Fruit & Vegetable Processing.

The fruits and vegetables at these levels are processed in the form of jams, jellies, squashes, syrups, murabbas, ketchup, pickles, canned/bottled fruits and vegetables, powders and spices.

Government Organisation

At Central Level : At the central level the union Government took a pioneering step by creating a separate Ministry of Food processing Industries in 1988. The new ministry was first of its kind in the developing countries with focus on integrating the interests of the farmer and the industry to promote better utilisation of agricultural commodities, greater values addition to rural produce, generation of massive employment in rural areas, enhancement of the net level of rural incomes and induction of modern technology in food processing. Another specific role of the ministry is to convert the large scale wastages of fruits and vegetables into useful food items, promote agro-based industrialisation in rural areas and help to absorb women and youth in gainful employment.

Ministry of food is also actively engaged in the promotion of fruit and vegetable processing through its Community Food and Nutrition Extension Unit which acts under food and nutrition board and exists at regional level.

At State Level : State government is also engaged in the activities of fruit and vegetable processing. In Haryana, the Haryana Agro Industries Corporation Ltd. (HAIC) has been identified as the nodal agency for promoting fruit and vegetable processing by encouraging investment by enterpreneurs and providing technical assistance. The plant is situated at Murthal in which a range of canned/bottled foods like jams, jellies, juices, tomato-ketchup etc. are produced. The products are marketed under the brand name 'DELICIA.

At District Level : At district level the field unit of the central food department acts in promotion of food processing by organising training in food and preservation to housewives, students and field personnel of women and child development, Health, Education, Agriculture and Rural development departments. These units are working for promotion of nutritional awareness through education and in providing service facilities to the needy community.

Non-Governmental Organisation

By the non-governmental organisations fruit and vegetable processing is done at different levels, i.e., by large scale units, small scale units, cottage units and at home level (Fig.2.2). **Large Scale Units :** The definition of large scale units as given in FPO is the factories with installed capacity of two metric tonnes of fruit products per day or having total annual production of more than 250 metric tonnes/annual value of production exceeding Rs. 1,00,000. The large scale units are engaged in preserving, conserving and extending the availability of food, providing training facilities, developing appropriate technologies and encouraging farmers to improve quality and productivity.

NON-GOVERNMENTAL ORGANISATIONS

Category	Area For	Raw Material Store	Finished Goods	Total	Height
Home Scale	25 sq mts	$10m^2$	$15m^2$	$50m^2$	10 Feet
Annual Prod. Up To 10 mt					
Cottage Scale	60 sq mts	$20m^2$	$40m^2$	$120m^2$	10 Feet
Annual Prod. Up To 50 mt					
Small Scale	150 sq mts	$50m^2$	$100m^2$	$300m^2$	14 Feet
Annual Prod. Up To 100 mt					
Large Scale	300 sq mts	$100 m^2$	$200m^2$	$600m^2$	14 Feet
Annual Prod. Up To 250 mt.					

Fig. 2.2 : Non-Governmental Organisation

Small Scale Units : The small scale units are those with an installed capacity of not more more than 250 metric tonnes per annum or with an annual value of production not exceeding Rs. 1,00,000. Small scale units are engaged only in preserving and conserving fruits and vegetables.

Cottage Units : The cottage scale units are those with an installed capacity of 50 metric tonnes annually or with an annual production not eceeding Rs 50,000.

Home Scale : At home scale the units are of two types, one with installed capacity of 1 tonne to 10 metric tonnes annually and the other ones are those which are performing fruit and vegetable processing activities for economic benefit of the family.

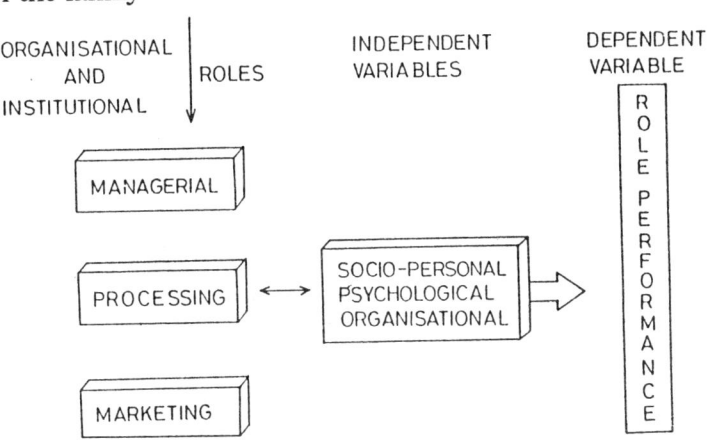

Fig. 2.3 : Role Analysis of Fruit and Vegetable Processing Organisations and Institutions.

Conceptual Model of Role Performance and Linkages of Fruit and Vegetable Processing Organisations and Institutions

The main objective of the conceptual framework developed in the present study is to provide a perspective or frame of reference which can facilitate analysis of role performance and linkages of fruit and vegetable processing organisations and institutions by rural families (Fig. 2.3).

From the model it was clear that three roles, i.e., Managerial, Processing and Marketing roles were expected to be performed by organisations at different levels. However, at the institutional level the roles would be slightly different for food processing being performed, i.e., commercial and economic purpòse. It was also conceptualised that these roles would be performed by organisations and institutions involved in fruit and vegetable processing. Socio-personal and psychological variables, infrastructural facilities which were presumed to have relationship with dependent variables,

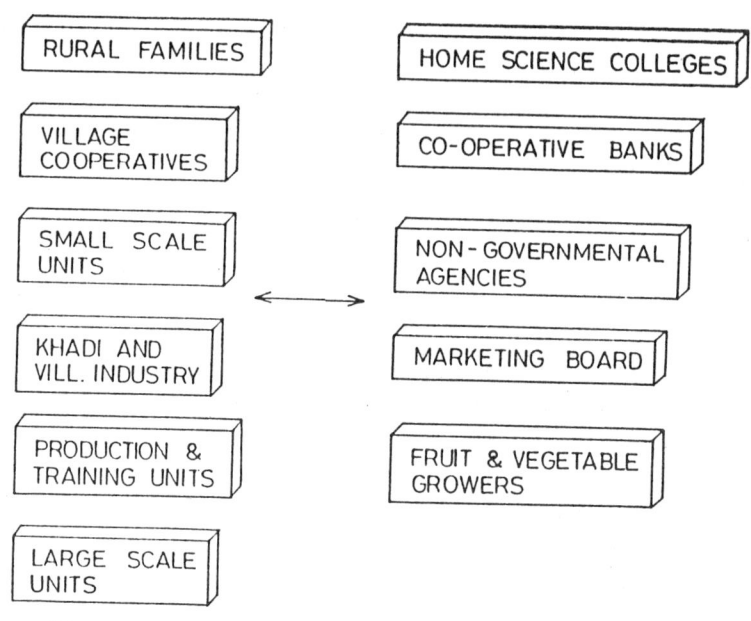

Fig. 2.4 : Linkages at Different Levels of Fruit and Vegetable Processing Organisations and Institutions.

i.e., role performance and linkages were included in the study. The study also conceptualised to examine the inter-organisational linkages (Fig. 2.4) with technical and financial institutes, horticultural and marketing boards and Government. This study also aims to unearth the influence of a set of independent variables with the role performance and linkages which would facilitate the role analysis of organisations and institutions working for fruit and vegetable processing.

3

REVIEW OF LITERATURE

A brief resume of past researches relevant to the present study has been incorporated in this chapter. The pertinent literature has been reviewed under the following sub-heads:

Role performance and linkages of fruit and vegetable processing organisations.

Appropriate fruit and vegetable processing Technologies for rural areas.

Factors influencing fruit and vegetable processing at different levels.

ROLE PERFORMANCE AND LINKAGES OF FRUIT AND VEGETABLE PROCESSING ORGANISATIONS

In this section, literature has been collected on various roles of organisations, i.e., production, production and training, training, research and development and linkages among them.

Role Performance of Organisations

Kurade (1981) stated that to popularise processing Department of Horticulture/Agriculture at State levels and Community canning centres can play a leading role by familiarising the housewives with processed fruit and vegetable products through lessons for preparation of squashes, jams, pickles etc. on a limited scale.

Mukhopadhyay (1981) observed that research Institutions must engage themselves in active research work to improve the farmer's productivity and to encourage him to grow those exotic varieties which could help the country to compete in the world markets.

Prasad (1981) quoted that one major role that an industry can play is providing producer-processor contacts and good rural-urban linkages, so essential for the growth and diversification of the rural economy.

Patil (1984) observed that industry has to play a major role in developing a suitable and appropriate technology to preserve perishable fruits and vegetables which, in turn, will help the farmers in securing remunerative prices for their products.

Anonymous (1986) reported that Indian Institute of Horticultural Research should involve technologists in food, fruit and vegetable preservation for selection of the projects before taking up programme of R & D. This will ensure projects of commercial importance.

Bagchi (1986) stated that Agriculture Minister urged the scientists for the development of new technologies in food processing particularly suited to rural women.

Geervani (1990) reported that home science colleges can play a major role in promoting fruit and vegetable processing in rural areas by :

(a) Identifying, developing and testing appropriate technologies at laboratory level.
(b) Conducting training for rural youth both women & men.
(c) Conducting demonstrations and compaigns to popularise new products.

Hindustan Times (1990) stated that Govt. provides grants for setting up of training centres and upgrading community canning centres and the scheme envisages to take up a package of extension, education and information programmes.

Mukherjee (1990) observed that Indian fruit and vegetable industry needs appropriate and low-cost technology in handling processing, packaging & preservation. This can be achieved easily with the help of Central Food Technological Research Institute (CFTRI) and the State Agricultural Universities (SAUS).

Xavier and Gopalaswamy (1990) stated that the Horticultural Research Stations need to play an important role in developing improved methods for quality control, grading, storage, processing and preservation of fruits and vegetables.

Anonymous (1991) reported that Ministry of Food processing act as prime force in creating an efficient and effective processing sector. Its role is to take initiative in mobilising cost-effective technologies for storage, processing and marketing of agricultural produce and also brings the power of modern technology and marketing techniques directly in aid of the farmers.

Economic Times (1992) reported that during the Eighth plan Government will set up more than 200 food processing & training centres in rural areas. These centres will impart training in all areas of food processing from production, marketing and quality control to overall management. Trainees will get practical experience of operating and managing a food processing unit at the small scale level.

Prahlad (1992) reported that Industry/University Co-operative Research Centres (IUCRCS) could be the beginning for new emerging trend for process development, product development, training facility of industry personnel and also establishing a new category of personnel for the requirement of the industry.

Raghunandan (1992) believed that financial institutions should play a leading role in the creation of infrastructural facilities through provision of loans to small sectors for encouraging fruit and vegetable processing in rural areas.

Rege (1992) observed that indigenous foods remained neglected in our research programmes. Even home science institutions did not undertake to optimise recipe of Indian culinary specialities.

Organisational Linkages

Mukhopadhyay (1981) suggested the idea of linking, growing and processing in the form of captive farms attached to processing units.

Tiwari (1981) reported that there are very few processing and manufacturing industries in the country which have such strong and direct linkages with the growth and development of our vast rural areas as the food preservation industries.

Jain (1982) in his study on the Inter-Organisational Linkages at village level reveal that there is absence of organic relationship due to social environment in which formal relationships among the individuals and groups have yet to become a way of life. No assistance, both financial and physical, was extended by any organisation to others.

Srivastava (1982) reported that organisations such as panchayat samiti, co-operative society and Agriculture Development trust have entered into operational linkages so as to avoid duplication.

Pandit (1983) stated that food processing industry has significant backward linkages. It can significantly contribute to the farm sector with better returns to the farmer by providing the incentives for improved varieties for horticultural produce and increasing overall employment.

Singh (1984) suggested that the industry should forge closer links with the farmers, agricultural universities and research bodies.

Joshi et al. (1989) suggested about the backward linkages of food processing industry that a shift in product-mix in favour of agro-based industries would have a favourable impact on income and employment generation.

Naik (1989) realised the importance of co-ordination and linkages among scientists and women so that their combined resources could render better services to women.

Chaudhary (1989) viewed that the linkages must be set up for transfer of technology by the research institutions like CFTRI, UDCT, Bomby and HBTI, Kanpur to KVI units free of cost to promote food - based industries for rural industrialisation.

Pandey (1989) said that presently there is no linkage between the processing industries and universities/institutes. New innovations are accordingly not being known to the industries. Processing industries do not have long term planning but take up processing on short term adhoc basis.

Devadas and Sujit (1990) reported that functional approach without proper linkages sometimes retard the enthusiasm of the weak participants to approach various agencies for technical and financial services.

Ghosal (1990) observed that magnitude of production loss can be prevented or atleast minimised by forgoing direct linkages with the processors or marketing outlets.

Ray (1990b) reported that because of so many hands involved in the development process, a policy may lose its identity by the time it reaches to the final stage.

Anonymous (1991) reported that a complete kind of linkage between the small and large sector is one in which the large sector provides every possible input to the small scale units to upgrade the variety of the product, the quality, the technology and the quality of the people in terms of training inputs.

Anonymous (1991) suggested that Khadi and village Industries Commission should form linkages with the state level boards to create a viable market network for products being promoted by them.

Rege (1991) observed that a close collaboration of industries, technologists, scientists, experts can create wonders in the promotion of food processing.

Roy (1991) revealed that research, teaching and extension are of prime importance in the growth and development of the processing of fruits and vegetables. However, scientific research without extension is not of much use as research finding must reach the actual user in order to have any impact. Therefore, systematic extension work must be taken up so that information regarding new products and methods is disseminated among the users.

Varde (1991) reported that principal missing link in Indian fruit and vegetable industry is not the lack of processing technology but the lack of an integrated technological approach encompassing all constituents from cultivator to consumer.

Sarin (1992) observed that few units which have been able to establish direct linkages with the farmers, productivity, quality and returns for the farmers have gone up, the farmers coming to share the prosperity of the industry.

Economic Times (1993) quoted that the industry must extend technological help to farmers to strengthen backward linkages as the raw material for the food processing sector comes from the agricultural sector. This would help lower the cost and improve availability of raw materials for the food processing sector, which contributes to high cost of finished goods.

APPROPRIATE FRUIT AND VEGETABLE PROCESSING TECHNOLOGIES FOR RURAL AREAS

Devadas (1980) reported that studies have shown that the keeping qualities of pickles can be extended by preserving in acid medium. Pickles preserved in oil and in 4 per cent acetic acid (vinegar), it was observed that acceptability of the pickles preserved by vinegar scored more by rural home makers.

Dhabade and Khedkar (1980) compared sun drying with cabinet drying of green mango slices and observed that the rate of drying was much faster in cabinet drying. In cabinet drying it took 10 hours to dry the pieces while 15 hours were required in case of sun drying.

Harigopal and Tonapi (1980) studied drying of grapes in solar dryer. They revealed that product exhibits marked improvement in colour and storage characteristics.

Harigopal and Tonapi (1981) reported that palm kernel is dried within 8 hours at an average cabinet temperature of 66°C. The product when sun-dried is white in colour, sweet in taste and soft and can be stored in glass bottles at room temperature for one year.

Kalra and Bhardwaj (1981) studied two models of solar dehydrator with mixed functions of direct and indirect dryers. Samples of mango slices, mango leather, green peas, okra and potato products were satisfactorily dehydrated. They further reported that the products were considered qualitatively superior to the open sun-dried products. The solar dhydrators are simple to fabricate and are well suited to rural conditions and small scale food processing industries.

Anonymous (1983) reported that wax emulsions given specific treatment under pressure were found to exhibit storage stability.

Maini et al. (1984) reported the physiological loss of weight in potatoes with an increase in the storage period. The desert cooling system helped more in reducing the physiological loss of weight by half of that of the samples stored at ambient conditions.

Maini et al. (1984) revealed that tray packed apples fetched higher over the conventionally packed fruits in reducing the bruising in apples. Tray packing is better as it avoids wrapping of individual fruits, prevents suffocation of fruits due to proper circulation of air among the fruits.

Maini et al. (1984) developed an efficient, simple, cheap, easily stretchable and foldable field type tubular dryer. The comparative performance of this solar dryer with other dryers for drying potato chips under winter conditions were studied and reported that this dryer is as efficient as a forced air fan operated dryer and better than a still type dryer or drying in the open. It took 6 hrs. to dry the chips in this dryer as compared to 9 hrs. in the open or in a still type dryer.

Maini et al. (1985) used "Forced indirect Solar air dryer" and "still type dryer" for the study and revealed that forced indirect solar air dryer was the most efficient one followed by drying in open on black polythene, still type dryer and plain polythene.

Anonymous (1986) conducted series of drying trials on the Galapagos islands in which three natural convection solar dryers were built and their performance compared with that of sun drying controls. All the dryers were more effective than traditional sun drying, reducing the drying time by one third. The quality of the product was higher as it was protected from insect and microbiological attack.

Khurdiya and Roy (1986) reported that in case of whole Ber the drying rate was faster in solar dryers with plain glass followed by amber glass, direct sun and solardryers with chimney. But the pattern was changed in case of potato slices, the drying rate of potato slices was faster in direct sun followed by solar dryer with plain glass, amber glass and chimney.

Roy and Khurdiya (1986) developed cooling zero energy input cool chambers using low cost, easily available materials. In these chambers the shelf life of leafy vegetables increased to 3 days from less than a day and for other vegetables the shelf life increased to 6 days comparred to 1-3 days. Apart from storing fresh fruits and vegetables these cool chambers can be utilised for increasing the storage life of processed fruit and vegetable products in non-corrosive containers.

Vaghani and Chundawat (1986) studied the sun drying of sapota fruits and reported that shelf life and the palatability depend upon the moisture content of the dried product. The lowest level of moisture 16% was recorded. From the point of view of taste, the dried product was acceptable upto 9 months with sugar syrup. Fruit slices treated with KMS (2%) had retained superior colour and appearance for 11 months.

Anonymous (1987) reported that storage life of fruit after picking with an inexpensive coating delays the ripening process. A simple-to-use kit, is suitable for apples, pears, cherries and invisible and tasteless sugar based film which is not toxic and edible. The increase in storage time varies according to the ripeness of the fruit when treated, but the natural life can typically be extended by four months for apples, 4-6 weeks for pears and 2-3 weeks for cherries and plums.

Malaviya et al. (1987) studied the effectiveness of drying chillies in open sun, cabinet dryer and natural convection solar dryer with and without chimney and reported that rate of drying in natural convection solar dryer was found to be faster in comparison to others.

Habibunnisa et al. (1988) studied the storage life of apples and oranges under evaporative cooling storage condition when treated with wax. they reported that evaporative cooling storage gave 6 times longer storage life for apples and 4 times longer storage life for oranges than at ambient conditions.

Maninck and Manimegali (1988) introduces Evapocooling chamber, a simple device to preserve the fruits and vegetables, extends the storage life of vegetables and reduces the weight loss in storage.

Mandhyan *et al.* (1988) studied dehydration of winter vegetables like peas, spinach, carrot and cabbage in the sun and in the solar cabinet dryer and drying constants were calculated. It was observed that the rate of moisture depletion in all the vegetables was high in the beginning and declined later. Reduction in the drying time was observed to be 15-20 percent when solar cabinet dryer was used in place of direct sun-drying.

Pawar *et al.* (1988) studied the comparison of drying in various solar dryers with that in mechanical and open air drying and indicated that the drying rate was the fastest in mechanical cabinet dryer followed by those in matrix bed air heater, rock type air heater (both solar dryers) and open air drying.

Sethi and Maini (1989) reported treatments like pre-cooling, waxing, disinfection, oiling, colouring, chemical treatments before the final grading and sizing for packing houses increases the shelf life.

Type of packing material also influences in reducing the weight loss. Packing of individual fruits either in paper, ventilated polythene bags, wax paper showed better quality and less shrivelling.

Zero energy cool chamber is best for retaining the freshness of fruits and vegetables during storage for a short period.

Processing of fruits and vegetables by sun drying, pickling, chemical preservation and lactic fermentation (vegetables especially) into more durable products will make them available over a longer period of time beyond harvest under the present socio-economic conditions prevailing in India.

Yadav *et al.* (1989) reported that to reduce the wilting and shrinkage losses, to lower the rate of respiration, to improve the appearance and to maintain the firmness of the product the vegetable waxing is done.

Chaturvedi (1990) suggested some low cost technologies for preservation of vegetables as products (fermentation, solar drying, use of preservatives) and preservation as whole fruits and vegetables (refrigeration and waxing).

Kumari Sarala (1990) reported that treatments like pre-cooling, waxing, disinfection, oiling, colouring, chemical treatment or fugicidal application before their final grading and sizing for packing will increase the shelf life of the commodity.

Kumari Sarala (1990) suggested solar dryers (chimney, amber coloured glass, home dryer) as best and simple low cost technology for dehydration of fruits and vegetables in the rural areas.

Kumari Sarala (1990) suggested zero Energy Cool chamber as best technology for retaining the freshness of the fruits and vegetables during storage. Small farmers can easily construct this chamber near their fields.

Mahadeviah *et al.* (1990) reported that plain indigenous E-25 and indigenous E-50 tinplate (general line quality) suitable for canning vegetables like potao, carrot, green beans and cabbage for a short storage life upto 3 months.

Mukherjee (1990) suggested that fermentation of vegetables, lactic fermented cauliflower, cabbage, green papaya, raddish, peas, beans etc. with salt will be of immense value to the farmers for home preservation during the peak production season.

Rai (1990) reported that higher the dry matter higher the product yield by solar dehydration. The yield of the flour was higher than that of the chips.

Jayaraman *et al.* (1991) revealed that although indirect sun-drying using solar cabinet dryer with plate collector took more time to dry and consequently resulted in slightly more shrinkage, the appearance, colour, flavour and organoleptic characteristics of the dried vegetables were much better than those of the direct sun-dried and comparable to those of hot air-dried. Direct exposure to sun resulted in significant loss of pigments.

Sethi *et al.* (1991) reported that whole white button mushrooms could be stored successfully without any spoilage upto 1year at room temperature in a chemical solution by simple method of steeping preservation. Chemical solution consisted of 2% salt, 2% sugar, 0.3% citric acid, 0.1% KMS and 1.0% ascorbic acid. Blanching before preservation has been recommended for long term storage for the enzyme inactivation to check colour deterioration during storage.

Chandel (1992) studied drying of fruits like wild apricot, apple, tomato and other vegetables as being practised in the region since the dried products are used during severe winters. However, the dust contamination leads to the poor quality of the dried product. The mud and wooden solar dryers of capacity 5-10 kg. have been installed under Desert Development project using U.V. resistant plastic film. The monitoring of the dryers has been carried out.

Malviya and Yadav (1992) studied comparative performance of convection solar dryer with and without chimney, with solar cabinet dryer and open sun drying. It was observed that the natural convection solar dryer without chimney, cabinet solar dryer and open sun drying, brought down the moisture content from 80% to 7%. The quality of dried chillies in natural convection dryer was better as there was no discolouration, charring or fungus growth, whereas in solar cabinet dryer discoloration and charring of the produce and in open sun drying deterioration of organoleptic quality were observed.

FACTORS INFLUENCING FRUIT AND VEGETABLE PROCESSING AT DIFFERENT LEVELS

Bennet and Tumin (1948) observed that role performance is not always in conformity with role definition.

According to Davis (1949), role is how an individual actually performs in a given position as dintinct from how he is supposed to perform. In simple terms, role performance is what the actors do as position occupants. The role behaviour is the same as the concept of role performance.

Shearer (1961) concluded from his study that women labour occupy a position of considerable role ambiguity. The roles they play are determined largely by the self image they bring to the job.

Kherdy and Sahay (1972) in their study reported that only two variables, i.e., attitude of the VLW towards the villagers and their knowledge regarding cropping pattern contributed significantly to the prediction of role performance of the village level workers.

Reddy and Mulay (1972) in their study revealed the following trend. In the progressive village, there was no agreement between leaders and non-leaders on the performance of various roles by leaders. In the less progressive village there was agreement between the leaders and non-leaders on the role performance indicating satisfaction among people of leader's role performance.

Chakravarty (1975) has observed that the role performance of rural women shows that work they do on the farm and in home contribute as much as half if not more to the economic development of the country.

Trivedi (1979) reported that the status of women workers in their families depends upon the kind of taboos associated with their role and duties within the family or outside when they have to work outside their home they are expected to earn and add to the family income. Change of status in family can occur on account of improvement in the educational level of family including that of concerned women.

Kaur (1981) reported that participation in farm as well as house related physical activities was more for daughter-in-law as compared to mothers-in-law and daughters of the household.

Kadam and Valunj (1982) reported that among socio-personal characteristics, education, social participation, socio-economic status and size of the family were significantly associated with the role performance of the gram panchayat members in village development activities. Among psychological characteristics, value orientation and cosmopolitanism were found to be significantly associated with role performance. Among general characteristics, working experience and adoption level are associated with the role performance.

Devi (1983) reported that the rural women belonging to low economic category performed more work in farm management, whereas those belonging to medium and high economic categories performed more work in house management. She further pointed out that caste and education were negatively and significantly associated whereas family size was positively and significantly associated with the farm and house management role performance of rural women.

Nikhade and Kitey (1984) reported in their study about the significant association of role performance of V.L.W. with their job experience and also stated that there was significant difference in the performance in respect of training received after service.

Bose and Sohal (1985) indicated that the record performance of the veterinary surgeons was average and the crucial variables that contribute to the prediction of job performance were perception of subordinate, total experience as VSs, inservice training and perception of internal control.

Sohi and Kherde (1985) concluded that the marital status, background, knowledge, attitude of Live Stock Assistants towards dairy farming, job satisfaction, perception of job significantly influenced the performance of Live Stock Assistants.

Kaur (1986) pointed out that with the development of agriculture, the extent of participation of rural women in various domestic activities had decreased as now they were in a position to own various time-saving household conveniences. With regard to their participation in farm activities, there did not seem to be any noticeable change except that for the management of cattle their role was becoming more of supervisory nature as for these tasks they were employing hired labour.

Khare *et al.* (1987) revealed that majority of young and old SMSs perform the role over and above the average level while adult SMSs perform below the average level.

It also unearthed that service experience, viz., greater length of service had no influence on the role performance of the SMSs.

Yadav and Azad (1987) reported that the backward and scheduled castes women are significant at 5 per cent level of significance and their respective percentages in agriculture are 65 and 55 of which 6 per cent women from backward and 8 per cent of scheduled castes prefer to go to agriculture side despite being educated. The inclination of scheduled castes women towards the agro-based industries is more in comparison to backward women.

Mishra *et al.* (1988) concluded that maintenance of daily diary dominated the habit of seeking scientific information and contacting farmers. They further reported that keeping close contacts with superior officer for getting advice dominates other important activities like helping farmers in obtaining loans and informing officers regarding the felt needs of farming community.

Sharma et al (1988) quoted in their study that educational qualification was related to the role performance. The role performance was found to increase with an increase in educational qualification.

Kunwar and Williams (1990) studied the discrepancies between the importance of tasks and their performance by junior technical assistants (JTA) as perceived by extension personnel. Extension personnel for importance of tasks indicate that all tasks could be included in the job description of JTA and tasks should be prioritized. The priority might differ from time to time, from region to region and, of course, from one system to another system.

Waris *et al.* (1990) reported that continuous guidance and regular supervision, effective cooperation and coordination, motivation and recognition for the work done will further enhance the role performance of Anganwadi Workers and increase their satisfaction with their job. Kadam (1991) reported by and large the role performance of contact farmers with respect to the roles considered was of medium to high level.

Rade et al. (1991) reported that the successful role performance of the contact farmers is an important indicator of success in the T and V system at grass root level. The role performance of the contact farmers was very satisfactory except feed-back role.

Bhople and Patki (1992) revealed that the farm women were involved in all types of activities but their maximum contribution was in pre-sowing, manuring, harvesting, grain-storage and marketing operations. The role performance of farm women coming from poor and backward castes with no formal education was found to be higher than that of the others.

Organisational Constraints

DGTD (1984) reported that the domestic constraint in the development of food processing are backward technology, lack of quality consciousness and higher prices of fruits and vegetables.

Krishna swamy (1987) highlights the practical problems faced by organisations engaged in food processing. There appears to be a shortage of technical personnel especially at lower level to operate modern processing machinery since adequate training facilities are not available.

Anomymous (1988) quoted that high cost and low demand plague food processing industry.

Vyas and Patel (1991) reported one of the main constraints in the development of industry is irregular supply of raw material.

Hindustan Times (1992) reported that lack of right management techniques, technologies, market information, research and development, information dissemination and poor quality of seed strains were some of the reasons for the poor performance in the processing sector.

4

RESEARCH METHODOLOGY

The objective of this chapter is to describe research methodology used in conducting the present study. Being exploratory in nature, the study was carried out through field surveys for better understanding of existing role analysis and linkages of fruit and vegetable processing organisations and institutions. The methodological steps, adopted in conducting the present investigation, included the locale of the study, research design, sample procedure, measurement of variables, identification of appropriate fruit and vegetable processing technologies, tools of data collection and data analysis procedure.

LOCALE OF THE STUDY

The present study has been conducted in Haryana State.

RESEARCH DESIGN

Exploratory research design was used to conduct the present study.

SAMPLING PROCEDURE

The study required selection of two different types of samples, i.e., respondents from both fruit and vegetable processing organisations and institutions. Hence, the sampling techniques for each has been explained hereafter, which is as follows:-

Selection of Districts

Out of 16 districts in Haryana, district Hisar, Panipat and Faridabad were purposively selected due to large number of fruit and vegetable processing organisations situated in these areas.

Selection of Organisations

Organisation has been operationalised as a collection of interacting and inter-dependent individual who would be working toward fruit and vegetable processing and whose relationship would be determined according to certain structure.

A list of various fruit and vegetable processing organisations in Haryana working at different level in the selected districts was obtained from the MInistry of Food, Krishi Bhawan, New Delhi (Annexure-1).

Initially it was planned to select only one big industry, however, later on it was decided to select one more to have more information. Hence, out of 7 large scale units, two units, namely, Pan Food Pvt. Ltd., Panipat and Pachranga International, Panipat were selected randomly. Similarly, two small scale units, namely, Murli Manohar Fruit and vegetable products, Panipat and Puja Achar factory, Hisar were selected randomly. There are only two Training units of Ministry of Food situated in Haryana. One in Hisar and another in Faridabad district and both were selected purposively. Similarly, there is only one Khadi Gramudyog unit and one co-operative unit which exists in Haryana and both are situated in Hisar district and were selected purposively for the study.

Selection of Institutions

Institution has been operationalised as a family having a

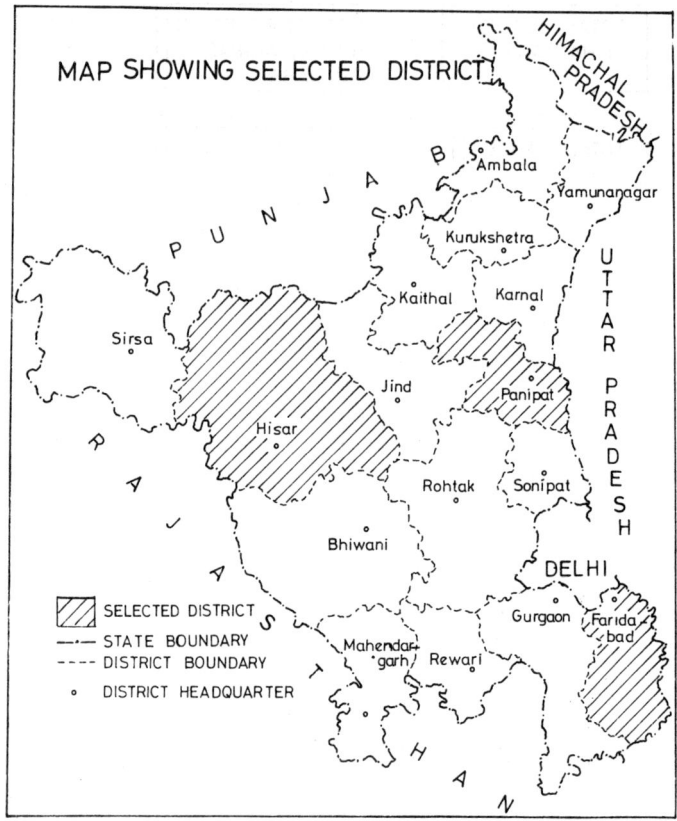

Fig. 4.1 : Map showing selected district.

social, educational or religious purpose working towards the fruit and vegetable processing.

For the selection of institutions, Snow ball method of investigation was followed, i.e., those families who were employed in the selected organisations and were also engaged in fruit and vegetable processing at home scale level whether for commercial purpose or for the economic benefit of the family, were selected. It was planned to select 100 families but could find 65 from the selected organisations/districts as shown in Fig. 4.2.

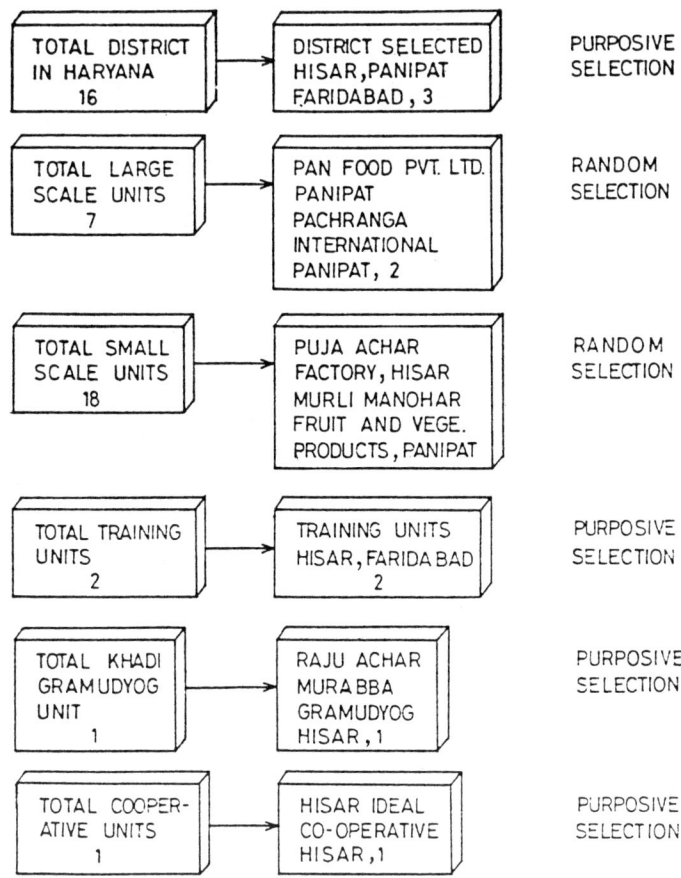

Fig. 4.2 : Sampling procedure for studying role
analysis and linkages of various
organisations at different levels

VARIABLES AND THEIR MEASUREMENT

Under this section the instruments/techniques to measure
the antecedent variables and dependent variables have been
incorporated both for the employees working in different
organisations and rural families separately.

Exployees : The background information of the employees
was collected mainly on the seven major aspects as shown
in Table 4.1.

Table 4.1. Variables for Organisation's Personnel

Variables	Scale used.developed
Education	Trivedi, 1963.
Total service experience at present post	Index (Varma, 1987)
Attitude towards job	Scale developed
Job satisfaction	Varma, 1987.
Organisational climate	Scale (Laharia, 1978).
Organisational commitment	Porter *et al.* (1974)
Professional commitment	Jauch *et al.* (1978)

Variables for Organisation's Personnel

Education : Education was operationalised as number of years of formal education attended by the respondent. Scores assigned were as follows :

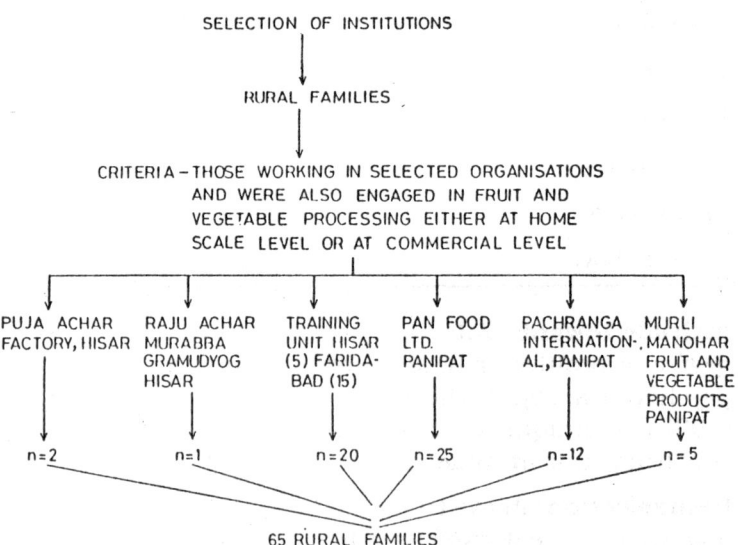

Fig. 4.3 : Selection of Rural Families.

Categories	Scores
Illiterate	0
Primary	1
Middle	2
High School	3
Technical/vocational education	4
Graduate	5
Post-graduate and above	6

Total service experience at present post : It refers to the total service experience on the present post in completed years in round numbers at the time of data collection and relative scores assigned to different categories are :

Categories	Scores
1-5 Years	1
6-10 Years	2
11-15 Years	3
16-20 Years	4
21-25 Years	5
26 and above	6

Attitude towards Job : Attitude has been defined as the degree of positive or negative effect associated with some psychological object (Thurstone, 1946). In this study, Likert (1932) technique of summated rating was used for construction of attitude scale.

Item selection : Above 15 statements comprised the content area which were derived from relevant literature.

Editing of statements : These statements were edited and modified and put to a preliminary scrutiny as per criteria suggested by Edwards (1957). The ambiguous and irrelevant statements were eliminated. After preliminary scrutiny 10 statements were selected.

Selection of Statements : The edited statements were presented to judges in order to determine their relevancy in context of practical practice. The complete set of statements were administered to 25 judges. The judges were requested to sort out the items by deciding the relevancy of the content of the item given and thereafter to determine the degree to which the statements would be indicative of the scope of the definition as per their judgement.

On computation of judge's response 80.0 per cent unanimity was observed among the judges which indicated a high degree of relevancy of the statements. All the statements were retained after modifications suggested by the judges for further analysis.

Analysis of statements : The item analysis was done with the application of summated rating method suggested by Likert, 1932. The items listed down on the basis of the consensus of the judges were then presented to 30 respondents of HAIC, Murthal who were selected at random and did not constitute the sample. The responses of rural families were obtained on a five point contiuum ranging from strongly agree to strongly disagree. Favourable statements were assigned the scores of 5,4,3,2 and 1 and for unfavourable statements scoring procedure was reversed. Minimum and maximum scores were worked out by multiplying number of respondents with obtainable minimum and maximum scores. In the present scale the minimum and maximum scores were 30 and 150 respectively. The respondents were arranged in descending order on the basis of obtained scores for a particular subject. At the top 25 percent respondents and at the bottom 25 per cent were taken as the criteria group in terms of which the individual statements were evaluated. Criteria groups were used for finding out the critical ratio of the items which is a measure of the extent to which a given statement differentiated between high and low groups.

The critical ratio for each of the statement was calculated and items with 't' values equal to or greater than 1.75 were selected for the final scale as persented in Annexure-II.

Reliability of the scale : The reliability of the scale was measured on a group of 20 non-sample rural families by test-retest method with an interval of 15 days. The reliability coefficient was found to be 0.69. It is clear from the figure that reliability coefficient was significantly high. Hence it was established that the scale developed to measure the attitude of employees of organisation was reliable.

Validity of the scale : The content validity of the attitude scale was fulfilled by the fact that items were collected from various pertinent sources. Further, the statements were finalised after judges opinion was sought, who were furnished with specific directions for making judgements. Therefore, it is confirmed that the scale developed to measure the attitude of rural families towards fruit and vegetable processing possessed the content validity.

Administration of the scale and scoring procedure : The scale was administered to the sampled respondents on a 5-point continuum ranging from strongly agree to strongly disagree. Their responses were recorded against each statement in the appropriate category. The scoring was done with the weightage of 5,4,3,2 and 1 for favourable statements and for unfavourable statements scoring procedure was reversed, ie., 1,2,3,4, and 5. In this way the total score for each statement was obtained.

Organisational climate : The dictionary meaning of the word climate is 'character of something'; Extending this meaning to the organisation we may say organisational climate as a character of an organisation. Scheinder and Snyder (1975) defined organisational climate as a summary perception which people have about an organisation. It is, then, 'a global impression of what the organisation is'. In the present study, organisational climate has been operationally defined as the state of fruit and vegetabe processing organisation's internal structure as perceived by the personnel. It was measured through the scale developed by Laharia (1978). The respondents were categorised on the obtained response into the following categories :

Categories	Range	Scores
Less satisfactory	(65-150)	1
Moderately satisfactory	(151-237)	2
Satisfactory	(238-325)	3

Organisational commitment : It was referred to the effective attachment to the goals and values of an organisation to one's role in relation to goals and values. This variable was operationalised by using the scale developed by Porter et al. (1974). It consists of 15 items both negative and positive statement. They were quantified on a five point continuum, ranging from strongly agree to strongly disagree. The respondents were categorised on the obtained response in the following categories :

Categories	Range	Scores
Low	(15-33)	1
Medium	(34-53)	2
High	(54-75)	3

Professional Commitment : Jauch et al. (1978) defined professional commitment as the psychological bond between individual and the programme. In the present study, it has been operationalised as the involvement in the activities of one's profession. This variable was measured and quantified in the study with the scale of Jauch et al. The respondents were categorised on the obtained response in the following categories :

Categories	Range	Scores
Less committed	(6-13)	1
Moderately committed	(14-22)	2
Highly committed	(23-30)	3

Job satisfaction : Smith (1955) defined job satisfaction as, "The employees judgement of how well his job on the whole is satisfying his various needs". According to Lacke (1976), job satisfaction is the pleasureable and emotional state resulting from the perception of one's job as fulfilling or allowing the fulfilment of one's important job values, provided these values are compatible with one's needs. Laharia (1978) remarked that job satisfaction is perception based and reflects internal psychological state of heart and head. It indicates the gap between one's expectation and achievements where lesser gap, higher is the job satisfaction. It is also multi-dimensional in nature. The multidimensional nature of Job satisfaction was also supported by Pestonjee (1973). In the present study, the job satisfction has been operationally defined as the degree of satisfaction and dissatisfaction of fruit and vegetable organisation's personnel on various important job related aspects. It was measured through the scale developed by Varma (1987).

Some information about the fruit and vegetable processing organisations such as their general, financial information, availability of machineries and raw material was also collected. These are mentioned in the schedule.

Variables for Rural Families

Personal Variables

Age : It was operationalised as the number of full years completed by respondents at the time of interview. Scores assigned to different age groups were as follows:

Categories	Scores
Young (Below 25 Years)	1
Middle (25-50 Years)	2
Old (Above 50 Years)	3

Family size : The size of family was operationally defined as the total members in the family and whether it is nuclear or joint. Trivedi's SES Scale (1963) was used to quantify the size of the family. Scores assigned were as follows :

Cotegories	Scores
Upto 5 members	1
5-10 members	2
Above 10 members	3

Education of respondent : It was operationalised as the number of years of formal schooling attended by respondents. Scores for different educational levels were given according to the socio-economic status scale of Trivedi (1963) and were as follows :

Categories	Scores
Illiterate	0
primary	1
Middle	2
High School	3
Technical/vocational education	4
Graduate	5
Post-graduate and above	6

Family Education Status : Education has been operationalised as the number of years of formal education attended by the family members of the respondents.

An index of education was developed by dividing the total number of years of formal education attended by the family members above 6 years of age with the total number of family members. The relevant information was obtained under following categories and scored as under :

Categories	Scores
Illiterate	0
Primary	1
Middle	2
High School	3
Technical/vocational education	4
Graduate	5
Post-gradute and above	6

Table 4.2. Variables for rural families

Variables	Scale/Index used
INDEPENDENT VARIABLES	
* Personal variables	
Age	Chronological age
Family size	Trivedi's SES Scale (1963)
Education of respondent	-do-
Family education status	Narwal (1982)
Status in family	Index developed
Social variables	
Caste	Trivedi's SES scale (1963)
Family Type	-do-
Urban Contact	Varma (1987)
Social participation	Trivedi's SES scale (1963)
Psychological variables	
Economic motivation	Moulick (1965)
Attitude	Scale developed
Risk orientation	Supe (1969)
Knowledge	Knowledge scale developed
Resource system variables	
Type of unit	Index developed
Machinery and equipment possession	-do-
Availability of technical/ managerial guidance	-do-
Availability of machinery and equipment	-do-
Availability of raw material	-do-
Constraints	-do-
DEPENDENT VARIABLES	
Role performance	Scale developed
Linkages	Index developed

On the basis of the procedure mentioned above family educational status was computed and family presented under low (0-1), medium (2-3) and high (3-4) categories.

Status in Family : It was operationalised as the status of female respondent in the family as daughter, daughter-in-law and mother-in-law and the scores given were 1,2 and 3 respectively.

It was also operationalised as the status of male respondent in the family as unmarried son, married son and father-in-law and the scores given were 1,2,and 3 respectively.

Social Variables

Caste : Caste refers to the class or distinct hierarchy orders of society. The operational measures of caste were taken from the socio-economic status scale developed by Trivedi (1963) and were classified into lower, middle and higher castes and relative scores were assigned. Scores assigned to different caste groups were :

Categories		Castes	Scores
Lower	(a)	Chamar, Bhangi, Doom	1
	(b)	Jhimar, Khati, Dhobi, Badhi	2
Middle	(c)	Lohar, Kumhar, Darji, Nai	3
	(d)	Baniya, Sonar, Ahir, Jolaha, Saini, Arora	4
High	(e)	Brahmin	5
	(f)	Jat, Rajput, Bishnoi	6

Family Type : Type of family means whether it is nuclear family or a joint family. A nuclear family is composed of members of only one person and includes minors and dependents. A joint family refers to one which is constituted by two or more brother's families. This variable was operationally measured by SES Scale (Trivedi, 1963) with the following scoring system :

Categories	Scores
Nuclear	1
Joint	2

Urban Contact : Urban contact refers to the degree of extent of visits of respondents to city and town for the last six months. The contacts were quantified on the basis of schedule developed. The score allotted took into account the frequency with which they had visited :

Frequency of visit	Scores
Visit to city/town	
(a) Never	0
(b) Seldom (between 6 months)	1
(c) Frequently	2

Social Participation : It refers to the degree of involvement of respondents in formal organisation and institution either as a member or as an office bearer. It was measured by adopting the following pattern of scoring based on Trivedi's Socio-economic status scale (1963) :

Categories	Scores
Nil	0
Member of one organisation	1
Member of more than two organisation	2
Office holder	3

Psychological Variables

Economic motivation : It is occupational success in terms of profit maximization and relative value an individual places on economic ends. In the present study economic motivation referred to one's inner desire to maximize production as well as profit from fruit and vegetable processing.

It was measured by the self-rating scale developed by Moulick (1965). The obtained scores ranged from 1 to 9.

The computation of economic motivation was done as under :

Categories	Scores
Most liked	2
Least liked	1

Attitude : According to Allport (1935)", attitude is a mental and neural state of readiness, organised through experience exerting a dynamic or directive influence upon the individual's response to all objects and situations with which it is related". In the present study, it has been operationalised as the attitude of institutions towards fruit and vegetable processing. It was developed through Likert (1932) technique.

Collection of Statements : Initially' 14 statements were collected or developed for fruit and vegetable processing attitude scale by consulting the available specialists in the concerned area.

Editing of statements : These statements were edited and modified and put to a preliminary scrutiny as per criteria suggested by Edwards (1957). The ambiguous and irrelevant statements were eliminated.

Selection of statements : The edited statements were presented to judges in order to determine their relevancy in context of practical practice. The complete set of statements was administered to 25 judges. The judges were requested to sort out the items by deciding the relevancy of the content of the item given and thereafter to determine the degree to which the statements would be indicative of the scope of the definition as per their judgement.

On computation of judges response 80.0 per cent unanimity was observed among the judges which indicated a high degree of relevancy of the statement. All statements were retained after modifications suggested by the judges for further analysis.

Analysis of Statements : The item analysis was done with the application of summated rating method suggested by Edwards (1969). The items listed down on the basis of the consensus of the judges were then presented to 30 respondents of Mittal Achar Factory who were selected at random and did not constitute the sample. The responses of rural families were obtained on a five point continuum ranging from strongly agree to strongly disagree. Favourable statements were assigned the scores of 5,4,3,2 and 1 for unfavourable statements scoring procedure was reversed. Minimum and maximum scores were worked out by multiplying number of respondents with obtainable minimum and maximum scores. In the present scales the minimum and maximum scores were 30 and 150 respectively. The respondents were arranged in descending order on the basis of obtained scores for a particular subject. At the top 25 per cent respondents and at the bottom 25 per cent were taken as the criteria group in terms of which the individual statements were evaluated. Criteria groups were used for finding out the critical ratio of the item which is a measure of the extent to which a given statement differentiated between high and low groups. The critical ratio for each of the statement was calculated and the items with 't' values equal to or greater than 1.75 were selected for the final scale as presented in Annexure-III.

Reliability of the scales : The reliability of the scales was measured on a group of 20 non-sample rural families by test-retest method with an interval of 15 days. The reliability co-efficient was found to be 83. It is clear from the figure that reliability co-efficient was significantly high. Hence it was established that the scale developed to measure the attitude of rural families was reliable.

Validity of the scale : The content validity of the attitude scale was fulfilled by the fact that items were collected from various pertinent sources. Further, the statements were finalised after judge's, opinion was sought who were furnished with specific directions for making judgements. Therefore, it is confirmed that the scale developed to measure the attitude of rural families towards fruit and vegetable possessed the content validity.

Administration of the scale and scoring procedure : The scale was administered to the sampled respondents on a 5 point continuum ranging from strongly agree to strongly disagree. Their responses were recorded against each statement in the appropriate category. The scoring was done with the weightage of 5, 4, 3, 2 and 1 for favourable statements and for unfavourable statement, scoring procedure was reversed, i.e., 1,2,3,4, and 5. In this way the total score for each statement was obtained.

Risk orientation : It refers to respondent's risk taking capacity and courage to face various types of problems encountered. This was measured by using risk preference scale developed by Supe (1969). The scale consisted of six statements. The responses were obtained under five point rating scale as strongly agree, agree, undecided, disagree and strongly disagree with scores of 5,4,3,2, and 1, respectively. The obtained scores were then categorised and scored as under :

Categories	Range	Scores
High	25-35	3
Medium.	15-25	2
Low	6-15	1

Knowledge : Knowledge has been operationalised in this study as " the amount of correct information possessed by rural families regarding various aspects of fruits and vegetable processing, i,e., management, processing and marketing. Knowledge was measured by 'Knowledge test' developed.

A standardised knowledge test as defined by Noll (1957) is one that has been carefully constructed by experts in the light of the acceptable objectives, procedures for administering, scoring and interpreting scores so that no matter who gives the test or where it may be given the results should be comparable. Keeping this in view, a standardised knowledge test about management, processing and marketing was developed by employing following scientific procedure :

(a) Item Selection

The content of the test consisted of questions called items. The important factor considered in collecting items for knowledge test was to cover fully the object measured by it. Relevant information about fruit and vegetable processing, its management and marketing was collected from the home science personnel, horticultural scientists and literature. Selection of items was based on following criteria :

1. Responses to statements promote thinking rather than memorisation.
2. They should differentiate the well informed women from less informed ones and have certain difficulty values.
3. The items included should cover all the areas of knowledge.

With these criteria in mind, items were prepared. Before editing of items they were subjected to expert scrutiny and then these items were framed in the objective form of questions to control bias if any.

(b) Item analysis

The analysis of a test yields three kinds of information, namely, index of the item difficulty, index of validity and index of discrimination. The first one indicates how difficult the item is while validity index shows how well the item scores and total score corresponds to each other. The third one indicates how items discriminate the well informed from the poorly informed respondents.

1. **Index of Item difficulty :** All the prepared items were administered randomly to 20 rural families who were quite different from the sample selected for the main study. Each respondent to whom the test was administered scored on the basis of score allotted-one for 'correct' or 'yes' answer and zero for 'incorrect' or 'no' answer. After computing the total scores obtained by each of the respondents on all items they were arranged in descending order. Item difficulty index was worked as the percentage of respondents answering a statement correctly. The item with P value ranging from 25 to 75 were considered for final selection of the knowledge test.

2. **Discrimination Index E 1/3 :** For this study, with E 1/3 value between 0.15 and 0.08 were considered for final format of the knowledge test. Since the discrimination index varies between 0 to 1, it was felt necessary to select the items with at least 0.15 to 0.30 discrimination index to have wider continuum of the item discrimination.

3. **Validity index :** The validity power of the item or its consistency with total scores in the test was worked by the point biserial correlation (Garret, 1965) and was calculated for each item separately. The significance of biserial correlation was tested with the help of 't' test (Guilford, 1956). The value of 't' with 38 degree of freedom (n-2) and 5 per cent level of significance were noted for coefficient of biserial correlation (Fisher and Yates tables).

The calculated value of difficulty index, discrimination index and validity index for all items have been incorporated in Annexure-IV.

Final selection of items : The item with difficulty index ranging from 25 to 75 (Garret, 1965) index of discrimination above 0.30 and significant biserial coefficient at 5 per cent were selected for knowledge test of rural families.

Reliability of the test : Reliability of the knowledge test was ascertained on a group of 20 rural families with the application of test-retest method and was administered to rural families, who were different from the main sample of the study at 15 days interval. The correlation coefficient was found to be 0.81 which indicated high reliability of the test.

Validity of the test : The validity of the test was ensured by biserial correlation. Only such items which depicted significant biserial coefficients were selected for inclusion in the knowledge test. Therefore, it confirmed the validity of the scale.

Administration of the test and scoring procedure : The rural families were asked to apply in the dichotomised categories, i.e., 'correct'-'incorrect', 'yes'-'no' and 'agree'-'disagree' for different statements. One score was allotted to the items of the knowledge reported correctly by the respondents. The aggregate knowledge scores were calculated on the basis of item of the knowledge test

Resource System Variables

Besides personal, social and psychological variables of the rural families, information about resource system variables was also collected.

Type of unit : It was operationalised as the level at which fruit and vegetable processing is being practised at institutional level. Categorisation and scoring of this variable was made in the following manner :

Categories	Scores
Commercial scale	2
Home-scale	1

Machinery and equipment possession : It was operationalised as the possession of machinery and equipment for the processing of fruits and vegetables by the rural family. The score of one was assigned to each machinery and equipment. It was categorised and scored as under:

Categories	Scores
Low possession (0-15)	1
Medium possession (16-30)	2
High possession (31-45)	3

Availability of technical/managerial guidance : The guidance and advice was provided by the agencies and technical institutes to the rural families while initiating, implementing and performing the tasks of fruit and vegetable processing. Categorization and scoring was done as under:

Categories	Scores
No guidance	0
Technical guidance	1
Managerial guidance	2

Availability of Machinery and equipment : It was operationalised as the need and availability of machinery and equipment required for fruit and vegetable processing. Scoring was done as under :

Categories	Scores
Not available	0
Available at selected places	1
Easily available	2

Availability of raw material : It was operationalised as the extent of availability of raw material needed by the institution for fruit and vegetable processing. It was categorised and scored as under :

Categories	Scores
Not available in time	0
Available at few places	1
Easily available	2

Constraints : It was operationalised as the impediments/ obstacles/barriers which hinder the efficient performance of organisations and institutions at various stages of fruit and vegetable processing. The exhaustive list of constraints was prepared in consultation with various persons and review of literature. Constraints were classified under broad categories of personal, social, psychological and infrastructural facilities for institutions and administrative staffing, communi- cational, linkages and coordination, budgeting, raw material and marketing for organisation. Their replies were asked simply on a dichotomy form, i.e., Yes/No.

Dependent Variables : Under this heading there are two variables, i.e., role performance of rural families in fruit and vegetable processing and their linkages with other organisations and institutions working for fruit and vegetable processing.

Role performance : Rizvi (1967) defined job performance as the manner and extent to which different jobs are performed in practical situation.

For the present study, role performance was conceptualized as various activities performed by rural families under following heading: Management, Processing (selecting, washing, peeling, chopping, blanching, preserving and packaging) and Marketing.

A list of 22 roles was identified. Rural families were directly questioned to indicate their roles on three point continuum ranging from always, seldom and never performed, which were assigned the scores of 2,1 and 0 respectively. Thus the maximum score one could attain was 44 whereas the minimum possible was 0. The respondents were categorised on the obtained responses into the following categories :

Categopries	Range	Scores
Good performance	30-44	3
Average performance	15-29	2
Poor performance	0-14	1

Linkage : "Linkage" means 'a series or system of links' according to Webster's dictionary. One of the meanings of "Link" is 'anything serving to connect or tie; all other meanings refer to "chain" in one way or the other. Linkage has been operationalised as the communication by fruit and vegetable processing organisations and institutions with technical institution, input organisation and credit organisations to facilitate coordinated movement for advancement of some productive purpose.

Linkages of rural families : The linkages of rural families with other rural families either of their own village or of other villages, with field functionaries of the state departments, with technical institutes and input agencies were studied with its mode of linkage. The responses were asked and scored as under:

Categories	Scores
Frequent	2
Seldom	1
Never	0

Inter-Organisational Linkages : Inter organisational linkages of the selected organisations were studied. Their linkages were studied with technical and financial institutions, field functionaries, with governmental and non-governmental organisations, with fruit and vegetable growers and with other agencies for technical, financial and marketing purposes with their mode of linkage. Their replies were asked simply on a dichotomy form, i.e., Yes/No.

IDENTIFICATION OF APPROPRIATE FRUIT & VEGETABLE PROCESSING TECHNOLOGY FOR DISSEMINATION IN RURAL AREAS

Appropriateness of the technology in the present context has been defined as the technlogy with low cost, low input, low risk type, rural bias, suitable for small scale application, use of local inputs and compatible with man's need for creativity.

To determine the appropriateness of the processing technology following methods were followed:

(i) Detailed description of all the technologies was furnished to judges and their extent of appropriateness was assessed as most appropriate, appropriate and least appropriate in context of rural families by assigning scores of 3,2 and 1 respectively.

(ii) Secondly, on the basis of research reports and secondary data certain recommended fruit and vegetable processing technologies were also identified.

(iii) Based on the results of above said two methods the processing technology which was judged/reported most appropriate for dissemination in rural areas was tested practically.

PREPARATION OF RESEARCH TOOLS

For data collection, two different types of structured interview schedules were prepared. One each for organisations and institutions, as discussed earlier. Before the actual administration of schedules, pre-testing was done at Haryana Agro Industries Corporation, Murthal, and for the other schedule, i.e., for institutions the pre-testing was done on 10 rural families of the Chabra Achar Factory, Hisar. Minor modifications were done to make the schedule more functional.

DATA COLLECTION

The data were collected personally with the help of structured schedule from organisations and institutions at their place of work and residence.

PROCESSING AND CODING OF DATA

The collected responses to each of the question were tabulated and coded on master sheets and appropriate tables with frequency distribution in specific categories were formulated.

ANALYSIS OF DATA

For the purpose of analysis and interpretation of results, different statistical tools employed are given below:

Per cent and mean were calculated to analyse the profile. Difficulty index, discrimination index, biserial correlation, reliability of co-efficients, critical ratio, correlation coefficients, multiple regression were also worked out with the formula given below:

(i) Index of item difficulty

$$P = \frac{ni}{Ni} \times 100$$

where,

P = Difficulty index in percentage of ith item.

ni = Number of women giving correct answer to ith item.

Ni = total number of women to whom ith item was administered.

(ii) Index of discrimination

$$E\ 1/3 = \frac{S_1 - S_2}{N/3}$$

where,

S_1 & S_2 = The frequencies of correct answers in the high and low groups respectively.

N = Total number of respondents in the item analysis samples.

(iii) Validity index

$$rp\ bis = \frac{M_p - M_q}{\sigma} \times \sqrt{\beta_q}$$

where ,

M_p = Mean score of respondent who gave correct answer to the item.

M_q = Mean scores of the respondents who gave incorrect answer to the item.

σ = Standard deviation of the entire sample.

p = Proportion of respondents giving correct answer to the item.

q = Proportion of respondents giving incorrect answer to the item.

(iv) Reliability of coefficients

$$r_{tt} = \frac{2\ rhh}{1 + r_{hh}}$$

where,

r_{tt} = reliability of the total test estimated from reliability of one of its halves.

r_{hh} = correlation between halves.

(v) Critical ratio

$$t = \frac{XH - XL}{\sqrt{\dfrac{\Sigma(XH - \overline{X}H)^2 + \Sigma(XL - \overline{X}L)^2}{n(n-1)}}}$$

where,

$\overline{X}_H$ = mean score on a given statement for the high group.

$\overline{X}_L$ = mean score on a given statement for the low group.

$\Sigma(X_H - \overline{X}_H)^2$ = The variance of distribution of responses of high group to the statements.

$\Sigma(X_L - \overline{X}_L)^2$ = The variaqnce of distribution of responses of low group to the statements.

n = Number of respondents in high and low groups respectively.

(vi) Correlation coeffient : It is the measure of relationships between dependent and independent variables and was calculated by employing following formula suggested by Snedecor and Cochran (1968).

$$r = \frac{\Sigma Xi\, Yi - \dfrac{\Sigma(X).\, \Sigma(Y)}{n}}{\left(\Sigma X^2_i - \dfrac{(\Sigma X)^2}{n}\right) x \left(\Sigma Y\, \Sigma y^2_i - \dfrac{(\Sigma Y)^2}{n}\right)}$$

where,

n = Number of respondents

$\Sigma X_i Y_i$ = Sum of products Xi and Yi

ΣX_i = Sum of Xi

ΣX^2_i = Sum of squares of Xi

ΣY^2_i = Sum of squares of Yi

i = n

(vii) **Multiple regression :** The relationship between each of the dependent variables Y_1, Y_2, Y_3, and Y_4 and significant variables in correlation was obtained by fitting the multiple regression equation (Snedecor and Cochran, 1968)

$$Y = a + b_1 X_1 + b_2 X_2 \text{—----— } + bk\ XK$$

where,

   y = dependent variable

   a = constant (intercept)

   b_1 = the slope of the first predictor

   k = total number of independent variables.

The regression coefficients 'bi' were tested by applying 't' test.

$$t = \frac{bi}{SE\ (bi)}$$

where,

SE (bi) = standard eror of ith regression coefficient.

The significance of R^2 was tested by applying the F test.

$$F = \frac{R^2/K}{(1-R^2)/(N-K-1)}$$

where,

   R^2 = Multiple regression

   K = Number of independent variables

   N = Number of respondents in the sample.

Case Study

The case study of eight organisations was done to have an intensive information about their functions, problems, finance, linkages, etc. The indepth interviews were done with the executive officers of these organisations. Some information was also collected in the questionnaire form.

5

RESULTS AND DISCUSSION

The results of the present research, derived through the use of prescribed methodology and standard tools mentioned earlier, have been presented in this chapter. These are in accordance with the said objectives and are described and discussed under different subheads as mentioned below :

- Role performance and linkages of fruit and vegetable processing organisations and institutions.
- Identification of appropriate fruit and vegetable processing technology for rural areas.
- Factors influencing fruit and vegetable processing at different levels.
- Strategy for promoting participation of women in fruit and vegetable processing.

Existing status of fruit and vegetable processing organisations in Haryana : The Fig 5.1 shows the categorywise break-up of fruit and vegetable processing organisations of the Haryana State. The data reveal that there were total 111 registered organisations working for fruit and vegetable processing in Haryana. They represented 7.20, 16.22, 18.02, 40.54, 18.02 percent large scale, small scale, cottage scale, home scale and relabelers type of fruit and vegetable processing organisations respectively.

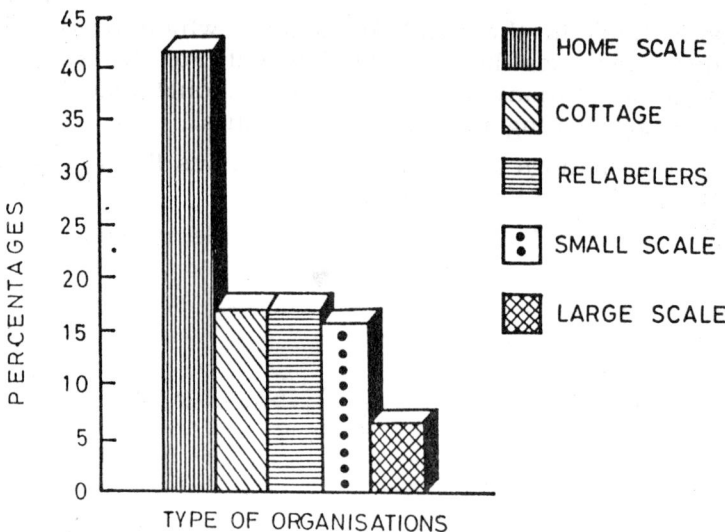

Fig. 5.1 : Category-wise break up of fruit and vegetable processing organisations in Haryana.

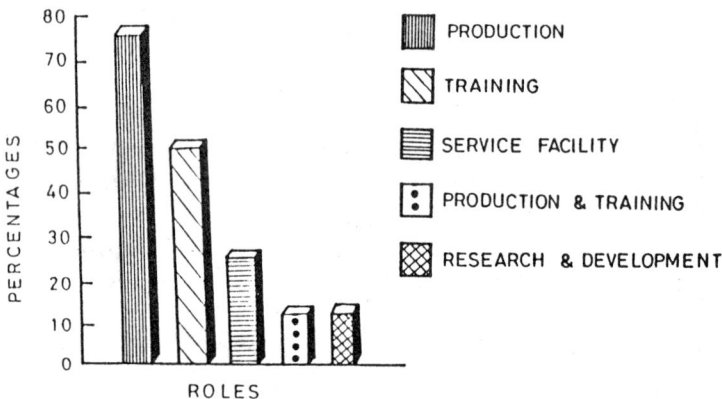

Fig. 5.2 : Different roles performed by fruit and vegetable processing organisations.

It can be construed that processing of fruits and vegetables was done at various levels of organisation. Further, it was also encouraging to note that 40.54 per cent of the organisations working for fruit and vegetable processing (home scale level) were registered. There were many such organisations in the rural areas also, which needed encouragement for the rural development. For this step, financial institutions and big industry should take the initiative by creating infrastructural facilities.

These findings are in tune with pervious findings of Bagchi (1986) and Ranghunandan (1992).

Roles performed by fruit and vegetable processing organisations : To study the role performance of selected organisations their main activities studied are presented hereafter. The data in Fig. 5.3 depict the main roles performed by various organisations. Data reveal that majority of the organisations (75.0%) were performing the role of production only, whereas 50.0 per cent were performing training roles followed by 25 per cent performing service facilities for common people. Only 12.5 per cent of the organisations were performing the role of research, development, production and training.

Thus, it can be inferred that organisations for fruit and vegetable processing at various levels were performing the role of research and development which is necessary for the technological advancement. These findings conform to the observation of Patil (1984), Anonymous (1990) and Mukherjee (1990).

Major products prepared by different organisations : Data in Table 5.1 reveal that in all the eight organisations pickle was the main product being prepared by them, whereas only the one organisation, namely, Pan Food Ltd. prepared various products like jams, jellies, syrups, squashes, murabba, ketchup, soups, dehydrated vegetables, dehydrated fruit and vegetable powder and canned products. However, the training centres under the Central Govt. also imparted training for preparation of jams, jellies, syrups, squashes, ketchup and pickles.

It can be inferred that organisations were preparing the age-old products, i.e., pickle and were not giving full attention to the new technologies and products when advancement in processing techniques is very fast.

Total manpower resources in the processing activities at various levels of organisation : The data in Table 5.2 show the manpower engaged in various organisations for processing activities. It is revealed that in all the organisations only 29 women were employed and that too for the manual activities like cleaning of spices, selection, grading, washing, peeling, chopping and blanching of fruits and vegetables, whereas 75 men were employed mostly for mechanical jobs in various fruits and vegetable processing organisations.

It can be construed that very less number of women were employed as compared to men and were not being trained for mechanical jobs.

Distribution of organisations personnel based on various attributes : Attitude of Organisational personnel towards job at different levels : Attitude has been defined as the degree of positive or negative effect associated with some psychological object (Thurstone, 1946). Attitude towards job was assessed at different levels, i.e., large scale units, small scale units, training units, khadi gramudyog and co-operative and the findings are presented below :

Attitude towards job at large scale units : The data in Table 5.3 reveal that majority of the personnel had moderately favourable attitude towards job in large units. Only very small percentage of personnel had unfavourable attitude.

Attitude towards job at small scale units : The data in Table 5.3 show that 50 per cent of the personnel were having moderately favourable attitude towards job in small scale units. And, 30 per cent of them had unfavourable attitude. Only 20 per cent of the personnel had favourable attitude.

Attitude towards job of training unit personnel : It was found that majority of the personnel had favourable attitude towards job. None of the personnel had unfavourable attitude.

Attitude towards job of khadi gramudyog personnel : It was observed that 85.33 per cent of the personnel had moderately favourable attitude towards job, whereas only 16.67 per cent had favourable attitude.

Attitude towards job of co-operative unit personnel : The data in Table 5.3 show that majority of the respondents (61.54%) had moderately favourable attitude towards job.

Table 5.1. Major products prepared at various levels of organisations.

PRODUCTS	Training Centres		Large Scale		Small Scale		Khadi Gramudyog Mandal	Cooperative
	Field units of central Govt. Department at Hisar	Faridabad	Pan Foods Ltd Panipat	Pachranga International Panipat	Murli Manohar Fruits & vegetable Products, Panipat	Puja Achar Factory Hisar	Raju Achar Murabba Gramudyog Hisar	The Hisar Ideal Co-operative Hisar
Jams	+	+	+	–	–	–	–	–
Jellies	+	+	+	–	–	–	–	.
Syrups	+	+	+	–	–	–	–	–
Squashes	+	+	+	–	–	–	–	–
Ketchup	+	+	+	–	–	–	–	–
Pickles	+	+	+	+	+	+	+	+
Soups	–	–	+	–	–	–	–	–
Dehydrated vegtable	–	–	+	–	–	–	–	–
Dehydrated vegtable powder	–	–	+	–	–	–	–	–
Dehydrated fruit powder	–	–	+	–	–	–	–	–
Canned Mushroom	–	–	+	–	–	–	–	–
Murabba	–	–	+	–	–	–	+	+

+ Prepared - Not prepared

Table 5.2. Total manpower resources in processing activities at various levels of organisations

Processing activities	Training Centres		Large Scale				Small Scale				Khadi Gramudyog Mandal		Co-Operative	
	Field units of central Govt. department at Hisar	Farida-bad	Pan Foods Ltd Panipat		Pachranga International Panipat		Murall Manohar Fruit & Veg. Products, Panipat		Puja Achar Factory Panipat		Raju Achar Muraba Gramudyog Hisar		The Hisar Ideal Co-operative Hisar	
	M	M	M	W	M	W	M	W	M	W	M	W	M	W
Cleaning of spices	–	–	–	5	–	2	–	2	2	–	2	3	–	2
Selection	–	–	–	5	–	3	–	2	2	–	2	3	3	3
Grading	–	–	–	8	–	2	–	3	5	–	3	2	2	2
Washing	–	–	.	15	.	3	.	3	2	.	2	3	2	3
Peeling	–	–	15	5	5	2	2	2	2	–	2	2	3	2
Chopping	–	–	20	5	15	3	5	2	5	–	5	2	2	3
Blanching	–	–	10	10	5	2	3	–	5	–	2	2	2	2
Preserving	–	–	35	–	10	2	2	2	5	–	5	3	5	2
Crushing	–	–	20	–	–	–	–	–	–	–	–	–	–	–
Pressing	–	–	10	–	–	–	–	–	–	–	–	–	–	–
Clarification	–	–	10	–	–	–	–	–	–	–	–	–	–	.
Filling	–	–	–	–	2	5	–	3	5	–	2	2	3	2
Sterilizing	–	–	20	–	5	–	4	–	.	–	–	–	–	–
Labelling	–	–	5	15	–	3	2	3	3	–	2	1	2	2
	–	–	40	15	15	5	5	3	5	0	5	3	5	3

M = Men;   W = Women

Thus, it can be inferred that majority of the personnel at different levels were having moderately favourable attitude towards their job. Attitude of the personnel towards their job affects their performance.

JOB SATISFACTION OF ORGANISATIONAL PERSONNEL AT DIFFERENT LEVELS

Job satisfaction has been defined as the employee's judgement of how well his job on the whole is satisfying his various needs (Smith, 1955). Job satisfaction or organisation's personnel on various important job-related aspects has been studied at different levels.

Job satisfaction of large scale unit personnel : The findings presented in Table 5.3 show that most of the personnel (48.7%) were moderately satisfied with their jobs in large scale units. And, 39.03 per cent of the respondents were satisfied with their jobs.

Job satisfaction of small scale unit personnel : The data in Table 5.3 depict that majority (50%) of the respondents were satisfied with their jobs in small scale units. only 20 per cent of them were not satisfied with their jobs in small scale units.

Job satisfaction of training units personnel : It was found that 50 per cent of the respondents had moderate level of job satisfaction, whereas rest 50 per cent of the respondents were satisfied with their jobs in training units.

Job satisfaction of khadi 'gramudyog' unit personnel : The data in Table 5.3 reveal that all (100%) of the respondents were moderately satisfied with their jobs in 'khadi-gramudyog' unit.

Job satisfaction of co-operative unit personnel : The data in Table 5.3 depict that 53.85 per cent of the personnel were satisfied wher as 46.15 per cent of them were moderately satisfied with their jobs. It can be construed that large per cent of the personnel were moderately satisfied with their job in large scale units and khadi gramudyog unit whereas in co-operative, training unit and small scale unit most of them were satisfied with their jobs.

Table 5.3. Distribution of organisational personnel at different levels based on attributes.

Attributes	Categories	Large scale n=41	Small scale n=10	Training unit n=8	Khadi gramudyog n=6	Co-operatives n=13
Attitude towards Job	Unfavourable (10-22)	2 (4.88)	3 (30.0)	–	–	–
	Moderately favourable (23-36)	22 (53.66)	5 (50.0)	3 (37.5)	5 (83.33)	8 (61.54)
	Favourable (37-50)	17 (41.46)	2 (20.0)	5 (62.5)	1 (16.67)	5 (38.46)
Job satisfaction	Not satisfied (20-45)	5 (12.19)	2 (20.0)	–	–	–
	Moderately satisfied (46-72)	20 (48.78)	3 (30.0)	4 (50.0)	6 (100.0)	6 (46.15)
	Satisfied (73-100)	16 (39.03)	5 (50.0)	4 (50.0)	–	7 (53.85)
Organisational Climate	Less satisfactory (65-150)	–	–	–	–	–
	Moderately satisfactory (151-237)	30 (73.17)	8 (80.0)	3 (37.5)	4 (66.67)	3 (23.08)
	satisfactory (238-325)	11 (26.83)	2 (20.0)	5 (62.5)	2 (33.33)	10 (76.92)
Organisational commitment	Low (15-33)	–	–	–	–	–
	Medium (34-53)	32 (78.05)	5 (50.0)	4 (50.0)	3 (50.0)	4 (30.77)
	High (54-75)	9 (21.95)	5 (50.0)	4 (50.0)	3 (50.0)	9 (69.23)
Professional commitment	Less committed (6-13)	2 (4.88)	–	–	–	–
	Moderately committed (14-22)	15 (36.59)	6 (60.0)	3 (37.5)	1 (16.67)	5 (38.46)
	Highly committed (23-30)	24 (58.53)	4 (40.0)	5 (62.5)	5 (83.33)	8 (61.54)

(Figures in parentheses indicate percentages)

Thus, the varied job satisfaction level of personnel greatly influences the performance and out-put of the unit.

ORGANISATIONAL CLIMATE AS VIEWED BY ORGANISATION'S PERSONNEL AT DIFFERENT LEVELS

"Organisational climate is the summary perception which the people have about an organisation" (Scheinder and Snyder, 1975). Organisational climate as viewed by organisation's personnel has been assessed at different levels and the findings have been presented hereafter:

Organisational climate as viewed by large scale units personnel : The data in Table 5.3 reveal that 73.17 per cent of the respondents viewed organisational climate as moderately satisfactory but 26.83 per cent of them viewed it to be satisfactory.

Organisational climate as viewed by small scale units personnel: It was observed that 80 per cent of the respondents in small units viewed organisational climate to be moderately satisfactory. A very small percentage of them viewed the organisational climate to be satisfactory in small scale units.

Organisational climate as viewed by training unit personnel : The data in Table 5.3 depict that majority of the training unit personnel viewed organisational climate to be satisfactory. And, 37.5 per cent of them viewed organisational climate to be moderately satisfactory.

Organisational climate as viewed by khadi gramudyog personnel : The data in Table 5.3 show that majority of the personnel viewed organisational climate as moderately satisfactory but 33.33 per cent of them viewed organisational climate to be satisfactory.

Organisational climate as viewed by co-operative personnel : It was observed that 76.92 per cent of the respondents in cooperative unit viewed organisational climate to be satisfactory, whereas rest of them viewed it to be moderately satisfactory.

It can be inferred that majority of the personnel at different levels of organisation viewed organisational climate to be moderately satisfying, whereas respondents of training units and co-operative viewed organisational climate as satisfactory.

The varied views of personnel at different levels regarding organisational climate might be due to their varied attitude and moderate level of job satisfaction.

Organisational commitment of Organisation's personnel at different levels : It has been defined as effective attachment to the goals and values of an organisation to one's role in relation to goals and values. Organisational commitment has been assessed at different levels, i.e., large scale, small scale, training unit, khadi gramudyog and cooperative unit and findings have been presented hereafter :

Organisational commitment of large scale unit personnel: The findings reveal that majority of the large scale unit personnel had moderate level of organisational commitment. And, 26.83 per cent of them had high level of organisational commitment.

Organisational commitment of small scale training & khadi gramudyog unit personnel : It was found that 50 per cent of small scale, training and khadi gramudyog unit personnel had medium level of organisational commitment. Rest 50 per cent of the respondents of all the three organisations were highly committed towards their organisation.

Organisational commitment of co-operative personnel : The data in Table 5.3 show that most (69.23%) co-operative personnel were highly committed to organisation. And, 30.77 per cent of the respondents had medium level of organisational commitment.

It can be construed that majority of the respondents of large scale units had moderate level of organisational commitment. As compared to other levels, co-operative personnel had high organisational commitment and fifty per cent of small scale, training and khadi gramudyog personnel had moderate to high level of organisational commitment. This trend might be due to facilities available to the staff at various levels of organisations.

Professional commitment of organisation's personnel at different levels : Jauch *et al.* (1978) defined professiona commitment as the psychological bond between individua and the organisation. Professional commitment has beer assessed at different levels, i.e, large scale, small scale training unit, khadi gramudyog and a co-operative unit.

Professional commitment of large scale unit personnel
The findings indicate that majority of large scale unit personne had high level of professional commitment and 36.59 pe cent of them were moderately committed towards thei profession.

Professional commitment of small scale unit personnel
The data in Table 5.3 show that most of the small scale uni personnel had moderate professional commitment and 4(per cent of the respondents were highly committed to thei profession.

Professional commitment of Training unit personnel : I was found that 62.5 per cent of the training unit personne were highly committed and 37.5 per cent of the respondent: were moderately committed for their profession.

Professional commitment of khadi gramudyog unit personnel : The findings reveal that 83.33 per cent of the respondent from khadi gramudyog unit had high level of commitment towards their profession, whereas 16.67 per cent were moderately committed.

Professional commitment of co-operative personnel : The findings indicate that majority (61.54%) of the personnel were highly committed to their profession and 38.46 per cent had moderate commitment.

Thus, it can be inferred that most of personnel in large scale, training unit, khadi gramudyog unit and co-operative unit had high level of professional commitment, whereas in small scale units majority of the personnel were moderately commited for their profession. This might be due to varied attitude and job satisfaction level of personnel engaged in fruit and vegetable processing at various levels of organisations. So, professional commitment of the personnel affects their peformance which then ultimately affects the production/output of the particular organisation.

Distribution of organisational personnel at different levels based on attributes : The data in Table 5.4 show that at various levels of organisations only 4.88 per cent of the respondents had technical education in the large scale organisation. However, 25 per cent of the training unit and 4.88 per cent of large scale unit personnel were educated upto post-graduate level.

The data further reveal that in khadi gramudyog and cooperative unit, 50.0 and 61.54 per cent respondents had 1.5 years of service experience, respectively. But 75.0 and 25.0 per cent of the training unit personnel had 11-15 years and 16-20 years of experience, respectively.

Constraints faced by organisations at different levels : The data in Table 5.5 depict the constraints faced by various organisations at different levels during processing of fruits and vegetables. As regards administrative constraints, all the constraints were faced by the governmental organisation, i.e., training units.

Further, it was observed that 75.0 per cent of the organisations faced the limitation of the staff as the major constraint.

Data also show that 75 per cent of the organisations faced lack of funds as one of the major constraints related to budget of the organisation.

Regarding raw material, 50.0 per cent of the organisations were facing lack of supply of raw material as their main constraint.

Majority of the organisations (75.0 per cent) encountered competition from established unit and 62.5 per cent faced difficulty in getting money from buyer after sale as the major constraints.

It can be inferred that in the private organisations there was no administrative constraint. It may be added that limitations of the lower staff, lack of funds, high prices of raw

Table 5.4. Distribution of organisational personnel at different levels based on attributes

Attributes	Categories	Large scale n=41	Small scale n=10	Training unit n=8	Khadi Gramudyog n=6	Cooperative n=13
Education	Illiterate	8 (19.51)	3 (30.0)	–	2 (33.33)	2 (15.38)
	Primary	5 (12.20)	–	–	–	3 (23.08)
	Middle	9 (21.95)	3 (30.0)	1 (12.5)	2 (33.33)	2 (15.38)
	High School	5 (12.20)	–	1 (12.5)	–	4 (30.78)
	Technical/ Vocational/education	2 (4.88)	–	–	–	–
	Graduate	10 (24.38)	4 (40.0)	4 (50.0)	2 (33.33)	2 (15.38)
	Post-graduate	2 (4.88)	–	2 (25.0)	–	–

(Contd.)

Attributes	Categories	Large scale n=41	Small scale n=10	Training unit n=8	Khadi Gramudyog n=6	Cooperative n=13
Service experience	1-5 Year	5 (12.19)	3 (30.0)	–	3 (50.0)	8 (61.54)
	6-10 Years	15 (36.59)	3 (30.0)	6 (75.0)	3 (50.0)	5 (38.46)
	11-15 Years	16 (39.03)	4 (40.0)	2 (25.0)	–	–
	16-20 Years	5 (12.19)	–	–	–	
	21-25 Years	–	–	–	–	–
	26 and above	–	–	–	–	–

(Figures in parentheses indicate percentages)

material, competition and difficulty in getting money from buyer after sales emerged as the major constraints for the performance of fruit and vegetable processing organisations. This contention corroborates the findings of DGTD (1984), Krishnaswamy (1987); Anonymous (1988); Vyas and Patel (1991).

Inter-Organisation Linkages : Fig 5.3 a,b,c show the inter-organisational linkages for technical, financial and marketing purposes. It also shows that one organisation, namely, Pan Food Pvt. Ltd. Panipat is only single organisation which had a wider linkages with other organisations and agencies. It had technical linkage with scientists of various institutes like Central Food Technological Research Institute (CFTRI), National Institute of Nutrition (NIN) and Haryana Agricultural University (HAU) through seminars, conferences and for providing training to their personnel in these institutions. The Pachranga International and Murli Manohar Fruit and Vegetable products had linkages with the Foreign agencies for export purposes because 85% and 50% of their products are being sent to other countries, respectively. Puja Achar Factory, Hisar and the Hisar Ideal co-operative had no linkages with other organisations and institutions. The training centre, Hisar and Faridabad had linkages with technical institutes like CFTRI, NIN and HAU for providing training to their employees in these institutes and they also call the specialists from these institutes for delivering lectures to their trainees. These organisations, i.e., training units also had linkages with other governmental departments like Social welfare, Health, Education and some non-governmental organisations. These units impart training to the anganwadi workers of ICDS programme and also to school chidren and members of other non-governmental organisations.

None of the organisations had linkage with fruit and vegetable growers, although the farmers of the nearby villages cultivated fruits and vegetables at their farms. It can be suggested that fruit and vegetable processing industry should form the linkages with growers and with technical institutes for the technological upgradation. This finding corrborates with the previous observations of Tiwari (1981); Singh

Table 5.5. Constraints faced by the fruit and vegetable processing organisations.

Sr. No.	Constraints	Large scale n=2	Small scale n=2	Training unit n=2	Khadi Gram-udyog n=1	Co-ope-rative n=1	N=8
1	2	3	4	5	6	7	8
1. Administrative Constraints							
	(a) Centralised planning	-	-	2	-	-	2 (25.0)
	(b) Delegation of responsibilities without power.	-	-	2	-	-	2 (25.0)
	(c) More of formalities and paper work	-	-	2	-	-	2 (25.0)
	(d) Lack of infrastructural facilities	-	-	2	-	-	2 (25.0)
	(e) Lack of feedback and monitoring	-	-	2	-	-	2 (25.0)
2.Staffing							
	(a) Inadequate staff	1	-	2	-	-	3 (37.5)
	(b) Incompetent staff	-	1	-	-	1	2 (25.0)
	(c) Limitations of lower staff	2	2	2	-	-	6 (75.0)
	(d) Lack of refresher course	-	2	-	-	-	2 (25.0)
3.Communicational							
	(a) Lack of communicational facilities	-	-	2	-	-	2 (25.0)

(Contd.)

1	2	3	4	5	6	7	8
	4.Linkages and coordination						
	(a) Lack of coordinated approach	-	-	2	-	-	2 (25.0)
	5.Budgeting						
	(a) Lack of funds	2	2	2	-	-	6 (75.0)
	(b) Limited travel grants	-	-	2	-	-	2 (25.0)
	6.Raw Material						
	(a) Lack of supply of raw material for uninterupted production	2	2	-	-	-	4 (50.0)
	(b) Availability of raw material at a higher price	2	2	-	-	-	4 (50.0)
	7.Marketing						
	(a) Competition from established units	2	2	-	2	2	6 (75.0)
	(b) Delayed disposal of produce	-	1	-	1	-	2 (25.0)
	(c) difficulty in getting money from buyer after sale	2	2	-	1	-	5 (62.5)
	8.Other						
	(a) Natural calamities	2	1	-	-	-	3 (37.5)
	(b) Government rules / policies	2	-	-	-	-	2 (25.0)

(Figures in parentheses indicate percentages)

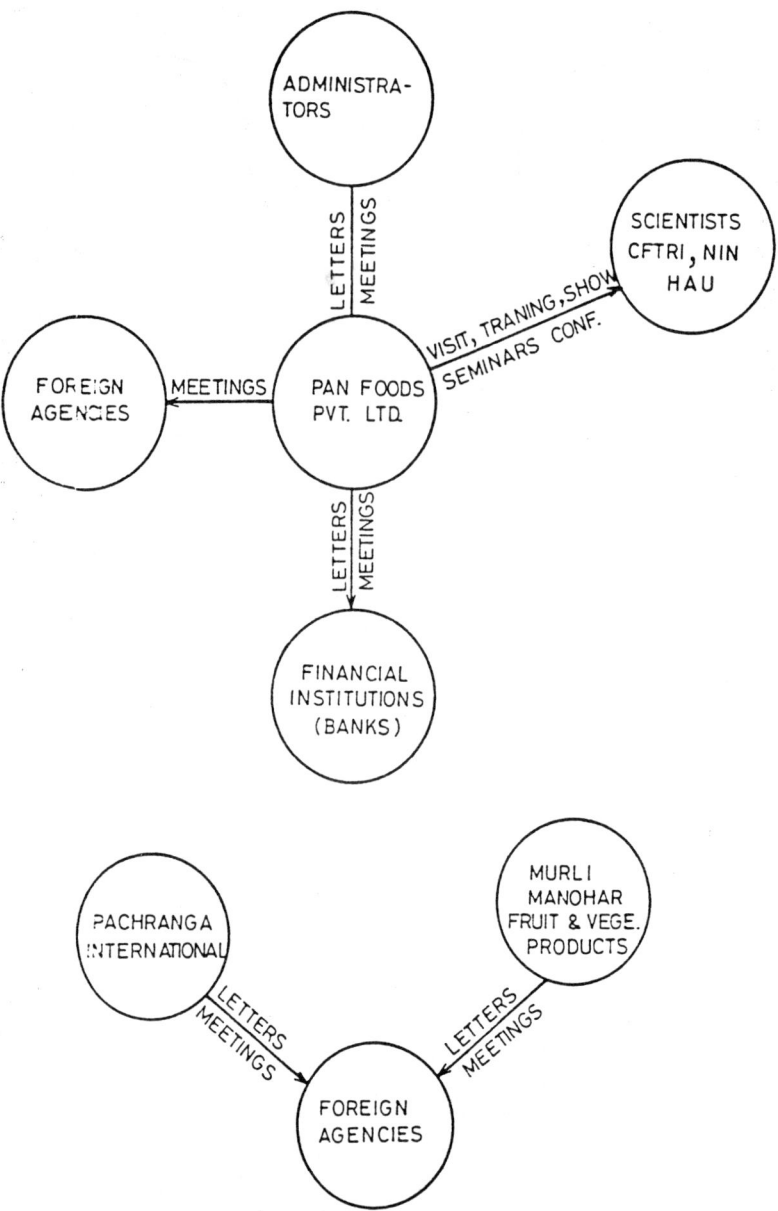

Fig. 5.3 (a) : Inter-Organisational Linkages

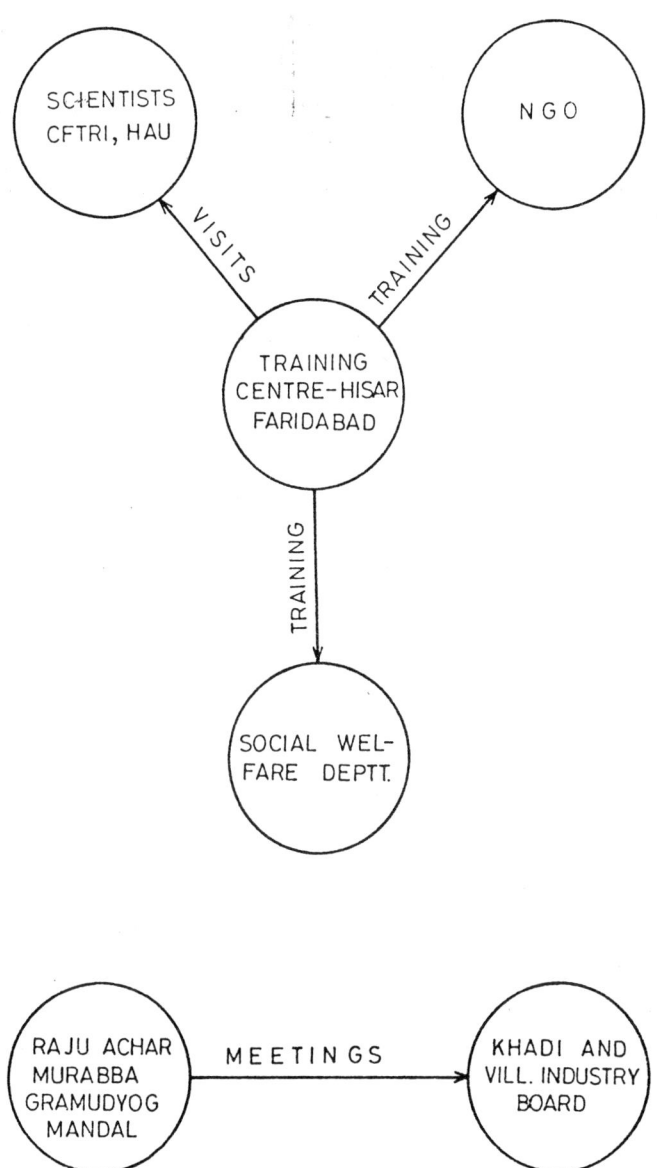

Fig. 5.3 (b & c) : Inter-Organisational Linkages

(1984); Naik (1989); Chaudhary (1989); Pandey (1989); Rege (1991); Roy (1991); Anonymous (1993).

Case Study, training Unit, Hisar : Case study of two training units was made, one at Faridabad and another at Hisar. Since organisational structure, infrastructural, financial and functional aspects of both the training units happened to be exactly similar, hence case study of Hisar training unit is presented.

Introduction : This training unit is named as " Community Food and Nutrition Extension Unit ", Central department of Food, Ministry of Food and civil supplies, Govt. of India, Krishi Bhawan, New Delhi.

This training field unit has been working for a number of years. This unit is located in a rented building 1624, Urban Estate, Hisar. The objectives of this unit are :

1. To impart education and training in :
    (a) Fruit and vegetable preservation.
    (b) Food and nutrition.
2. To provide service facilities for the preservation of fruits and vegetables.
3. Promotion of nutritional awareness through education.

However, the major role of this organisation is training in fruit and vegetable preservation.

The performance of roles of the unit is given here-under:

Roles : The national network for training consists of 33 centres, out of which two centres exist in Haryana State. The centre organises training, particularly for both urban and rural housewives. Training for preservation of fruits and vegetables is organised with an objective of educating them to start this work at home scale. For each training, a trainee has to pay a nominal fee of Rs. 5/-. The duration of the training is generally two weeks. The number of trainees for one training course is 25. Training is imparted through demonstrations, lectures, discussion, film and slide shows and exhibition. During a training course instructions are imparted to prepare the following products :

    (a) Jam, jelly and marmalade;
    (b) Preserves and candies;

(c) Fruit syrup;

(d) Squashes;

(e) Chutneys;

(f) Pickles;

(g) Ketchup and sauce etc.

For the preparation of above mentioned products, the trainees are required to bring their own raw-materials and they carry home their prepared products. The trainees are educated on different aspects of food and nutrition as well, so that they can bring about improvement in the nutritional status of the foods consumed in their home. Also, for children, instant weaning mixtures from roasted grains and pulses are got prepared. At the end of each training, certificate of participation is awarded to each trainee.

Trainings are also organised at the request of women of tribal areas, scheduled caste, backward classes and for the industrial families.

An examination of the training unit, Hisar revealed that yearly a large number of trainees are educated. In substantiation of this statement the requisite figures are presented below :

Year	No. of trainees
1986-1987	1001
1987-1988	1598
1988-1989	1178
1989-1990	741
1990-1991	1560
1991-1992	658

Organisational resources : The covered area of the building is 1984 sq. mts. The training unit is manned by :

1. One demonstration officer/officer-in-charge (M.Sc., diploma in food technology),

2. Technical Assistant (M.Sc. Hort),

3. Lab Assistant (Intermediate) and

4. Lab Attendant (Matric)

There are well defined roles of the different functionaries in the unit. The roles of Demonstration officer are :

(i) Planning and organising training and processing activities.

(ii) Liaison work with State Government and voluntary organisations.

(iii) Office administration.

The roles of Technical assistant are :

(i) Organising and conducting training programme.

(ii) Supervising processing activities.

(iii) Assisting in office administration.

The Lab assistant and Lab attendant help in the processing activities during training.

Input supplies : Raw materials are generally brought by the trainees themselves. However, some materials like chemicals, small tools and equipments are purchased as per government rules and regulations. By and large marketing of the products prepared is not required to be handled by the staff of the training unit.

Linkages : The personnel of training unit have linkages with technical institutes like home science collegtes, state department of agriculture and departments of horticulture and vegetables of state Agricultural Universities/Institutes for drawing resource persons to give lectures and demonstrations to the trainees of various courses.

For inservice training, the personnel are deputed for short training in various institutes like Central Food Technological Research Institute (CFTRI), National Institute of Nutrition (NIN).

Food and nutrition extension unit also provides training to field personnel of the departments of Women and Child Development, Health, Education, Agriculture and Rural development, Government Schools and Colleges.

Pachranga International, Panipat (Large Scale Unit) : On personal discussion with Sh. Asha nand Dhingra (owner of the organisation), it came to light that even before partition of India in 1947 he had been carrying on food processing

Training imparted to rural women in fruit and vegetable processing at training unit.

Plate 1 : Teaching method of preparing jam

Plate 2 : Women Preparing mixed fruit jam

work in the erstwhile Punjab (now in Pakistan). In 1947, when he came to India, he carried on the food processing work in U.P and Punjab and ultimately shifted to Panipat in Haryana State in 1950. Formelrly, pachranga International was started by five brothers but now it is the organisation of Sh. Asha Nand Dhingra and his sons. Pachranga International is located at G.T Road, I.B. Leiah School Lane, Panipat. it was first started with Rs. 25,000/- but now the capital investment is about 1 crore. While looking behind hesitation in discussing actual facts and figures about the organisation, it is very much likely that capital investment may be quite high.

The trade mark of Pachranga International is PIP. It is very prominently written that " Taste is better than rate", ask by name PIP.

Roles : The major role of Pachranga International, Panipat is only production. The organisation does not deal either with training or research. The organisation prepares pickles of different type, i.e., Pachranga International mixed pickle, mango unpeeled, lime and green chillies, mango peeled, lime spiced, turnip and cauliflower (sweet), ginger lime, green chillies, tenti/dela and red chillies stuffed. The speciality of this organisation is that all the products are prepared in mustard oil. From 15-7-92 they have added some more products like mango chutney, mixed fruit chutney, mushroom pickle, garlic pickle, zamikand pickle, amla pickle, Bhiss and lime sweet pickle.

All these products are packed in bottles (400 gms); machine sealed tins (800 gms and 2.25 kg); hand sealed tins (15kg) and plastic sealed jars (1,2,5 and 12.5 kg). The organisation proudly displays that all these products are prepared in Punjabi tradition to deliver a good taste. It may be mentioned that Pachranga International has been awarded many times by State Govt. of Haryana and Govt. of India for its quality, variety, flavour and taste. This organisation has earned distinction in getting export award as well.

Organisational Resources : The area of the space used by Pachranga International is 1300 sq. yds. Almost the whole area is covered. The organisation possesses all types of requisite equipment ranging in cost more than Rs. 5000/- to more than Rs. 50,000/- The equipment is procured from local market, nearby cities, places outside district and outside state. It was told that no difficulty is encountered in the procurement of equipment. By and large the machines are locally made.

The organisation is manned by father and his four sons. Individually, they look after various roles like management, production, purchase, sales and maintenance of accounts. Besides, 20 males and 15 females have been engaged to carry out various physical works. The male labour is literate and somewhat skilled while female labour is illiterate and unskilled. Female labour generally attends to cleaning of spices, washing of bottles, jars, fruits, peeling, labelling etc.

Input Supply : Raw-material for the preparation of different products is purchased from local market, nearby cities, places outside district, outside state and outside country.

Fruits and vegetables are purchased from local market through auction, whereas other inputs like machines and equipments are purchased direct from the company and through retailers thereof.

Marketing : The 85 per cent of the products are exported through foreign agencies to various countries such as Middle east (Gulf), Singapore, U.K., U.S.A., Canada, Australia, Germany and Japan. The rest of the products are sold through retailers and wholesalers within the Haryana State and the country.

Linkages : The organisation does not deal either with research or training and deals with production only. The largest part of the products is exported through foreign agencies and only a small part of the products is sold through retailers and wholesalers within the state and the country. All this clearly speaks of limited linkages of the organisation with foreign agencies, local retailers and wholesalers.

Pan Food LTD. Panipat (Large Scale Unit) : The Pan Foods Ltd., Panipat came into existence about two decades ago. Prior to becoming a large scale unit, the food processing work was being done on small scale. But at present this is a quite big food processing unit. It is located at G.T. Road, Panipat about 5 kilometers on Delhi side. The establishment has

**At various levels of
organisations women performing manual works**

Plate 3 : Women Cleaning spices

Plate 4 : Women washing bottles

been finaced by Punjab National Bank, Initial capital investment was told to be Rs. 1 crore and the running capital to the extent of Rs. 1 crore. Approximate sale proceeds of products per year were told to be Rs. 3 crores excluding export.

Roles : Pan Food Ltd. is a very large organisation and deals with all the three functions, i.e., production, training and research. However, the latter two functions, i.e., training and research are carried out on a limited scale.

The processed products comprise dehydrated peas, tomoto ketchup, all types of jam, and sweet ketchup,all soups, gulab jamun mix, dehydrated vegetable powder, dehydrated fruit powder and canned mushroom. The yearly respective production of these products was 50 tonnes, 400 tonnes, 400 tonnes, 50 tonnes, 10 tonnes, 30 tonnes, 100 tonnes, 50 tonnes and 100 tonnes. Also pickles are prepared for export only.

Processing : The products are prepared by using different methods of processing like canning, drying, use of chemical and other preservatives, filteration and bottling.

The organisation does not carry out any regular training programme. Through discussion it was revealed that one or two students from National Dairying Research Institute, Karnal come for training for 1 month duration.

To me it seemed to be a matter of interpretation perhaps the students of NDRI were sometime deputed to this organisation for familiarisation of the products and preparation as a component of their course work or research work.

The incharge of the organisation was not very clear in explaining the training role of the organisation because the clients did not belong to urban or rural families and there were only a few students of only one nearby institute.

With regard to research and development, no specific programme was told, however, it was told that some new products are developed and prepared. The list of products, of course, supported the veracity of the statement.

Organisational Resources : Total area of the Pan Food Ltd. was five acres. Out of this, 60 per cent area was the covered area including workshop and offices etc. The equipment included all types of machines costing more than Rs. 5,000/- to more than RS. 1 lakh. The machines and equipment were purchased from local market, nearby city, far off city, places outside district, outside state and outside country.

The manpower included one factory manager, one production manager, one quality control officer, one assistant production manager, one shift incharge, one shift supervisor, one assistant supervisor, one operator, one assistant operator, one chemist and labourers. As far as specific role of an incumbent is concerned, it is very much clear from the designation itself. All the above-mentioned functionaries constituted the regular trained personnel. Only the labour was engaged on daily wages. The number of labourers engaged per day depended upon the seasonal quantum of work lined up for the production of different products. Further, it may be mentioned that 40 male & 15 female labourers were engaged. Males were literate and somewhat skilled while females were mostly illiterate and unskilled.

INPUT SUPPLY

The raw-material was purchased from local market, nearby market, far off market, places outside country. With regard to procurement of raw material, no constraints were encountered.

Fruits and vegetables are generally purchased from market through auction, whereas machines, equipmemts, chemicals and packaging tins are purchased in the same manner as mentioned in the above para.

Marketing : Marketing of almost all the products is done within the state and in the country through retailers and wholesalers. Only pickles are exported. Based upon extensive marketing of the products through out the country, it can be safely stated that their products are quite popular.

LINKAGES

This organisation seems to have established extensive network of linkages. It was told that they have linkages with State Agricultural University/Institute, State Deptt. of

Plate 5 : Women separating labels for pasting

Plate 6 : Labelling tins

Agriculture, State Deptt. of Horticulture, Financial institutes and Foreign agencies. Due to qualified staff of the organisation there is adequate participation in relevant seminars held by related organisations.

MURLI MANOHAR FRUIT AND VEGETABLE PRODUCTS, PANIPAT (SMALL SCALE UNIT)

Murli Manohar fruit and vegetable product is located in House No. 331, Model Town, Panipat. This organistion has developed over the years. Infact, this happened to be a part of Pachranga International. therefore, food processing work was being done even before partition of the country in 1947 and before shifting to Panipat in 1950 this work was carried out at several places in U.P. and Punjab. In 1950 Murli Manohar fruit and vegetable product separated from the parent organisation, i.e., Pachranga international and started the food processing work on a small scale. Even presently this is a small scale unit.

ROLES

The Major role of Murli Manohar Fruit and Vegetable product, panipat, is only production. The organisation does not deal either with training or research. The organisation prepares pickles of different type, i.e., mixed vegetable pickle, mango, lemon, chillies, lotus stem, dela, ginger, turnip, cauliflower, carrot and garlic. The total yearly production of these products was about 2,000,00 Kg.

ORGANISATIONAL RESOURCES

The plinth area of the building is 3500 sq ft. The covered area is 600 sq. ft. Being a small unit, it possesses four machines costing Rs. 5000/- each and one machine costing Rs. 20,000/-. The equipment is procured from Delhi.

The organisation is manned by father and son, five males and three females. The males are literate and some what skilled while females are illiterate and unskilled.

INPUT SUPPLY

Raw Material is purchased locally and from Delhi.

**Men performing mechanical jobs at
different levels of organisations.**

Plate 7 : Sealing cans

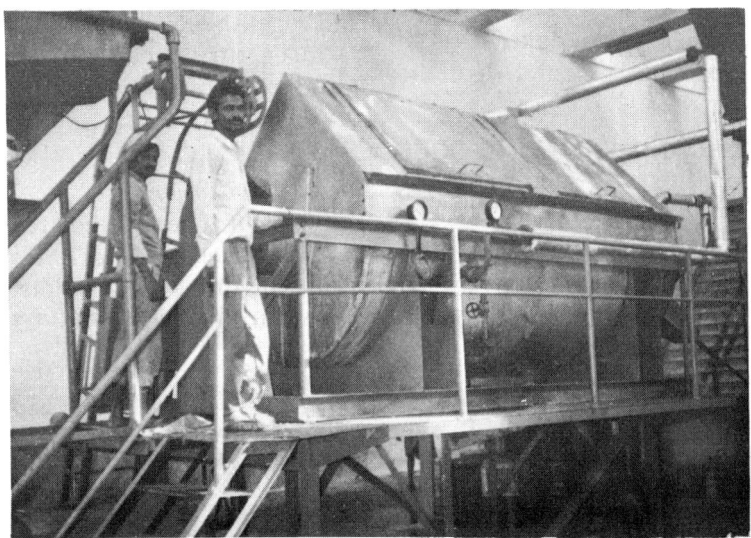

Plate 8 : Operating dehydrating mechaine

MARKETING

About 50 per cent of the pickles are exported through foreign agencies to various countries like middle east (Gulf), Singapore, U.K., Canada, Australia, Japan and Germany. The rest of the products are sold largely through retailers and wholesalers within the state and outside the state.

LINKAGES

Linkages are very limited for sale purposes only, i.e., with the foreign agencies and some retailers and wholesalers.

Puja Achar Factory, Hisar (Small Scale Unit)

Puja Achar Factory was started in the year 1987. It is located in House No. 83, Near New Sabzi Mandi. It started as a small unit and presently has the same status. The work was started with initial investment of Rs. 10,000/- which was personal. Yearly sale proceeds amount to Rs. 40,000/- while the net profit per year is about Rs. 10,000/-.

ROLES

Production of pickles is the only activity of this unit. It does not deal with any other aspect like training, research and development. The pickles are prepared from mango, lemon, chillies, tenti/dela, lotus stem (Bhain), amla, cauliflower, carrot and ginger. But all these pickles are produced in small quantity.

ORGANISATIONAL RESOURCES

Plinth area of the building is 1400 sq. ft. while production area is about 280 sq. ft.

This organisation possesses only one tin sealing machine costing about Rs. 5,000/-. Machines and equipments are purchased locally and from Delhi.

Organisation is run by members of the family, i.e., father and his sons. Whenever needed a couple of labourers are engaged seasonally.

Input Supply : Raw material is purchased locally and from places outside the state.

Marketing : the products are sold locally within the state.

Linkage : there is no linkage except for sale purposes with retailers and wholesalers.

Raju Achar Murabba Gramudyog Mandal, Hisar (Khadi Gramudyog Unit) : Raju Achar Murabba Gramudyog Mandal was started in the year 1989. This is located in house No.3, Sant Nagar, Hisar. It was started with initial investment capital of Rs. 10,000/-. The running capital is presently Rs. 72,000/-. Loan was obtained from Khadi board. Actually, this is a Khadi Board- registered organisation. Approximate sale proceeds per year amount to Rs. 2 lakh while yearly net profit is Rs. 60,000/-.

Roles : The major role of the organisation is production and to some extent training as well. The processed products are of lemon, chillies, mango, teent/dela, ginger and garlic. The yearly total production of all the products amounts to 10 tonnes. The organisation has imparted training to only three persons for the duration of six months.

Organisational Resources : The plinth area of the building is 90 sq. yds. while the workshop area is 50 sq. yds.

It possesses one machine worth more than Rs. 10,000. Machines and equipments were purchased locally and from Delhi.

The organisation is actually khadi board registered society consisting of seven members. One of them is president, another secretary and third is treasurer while others are as members. The president calls meetings for taking decisions and is responsibile for sanctions for different transactions. The secretary acts as manager and the treasurer looks after financial affairs. The unskilled labourers consist of 5 males and 3 females.

INPUT SUPPLY

The raw material is purchased from the local market and Delhi through auction.

The products are being sold locally and within the state through retailers.

LINKAGES

Linkage of this organisation is with Khadi board for financial assistance and with retailers for sale purposes.

The Hisar Ideal Fruit And Vegetable Products (Cooperative Unit) : This cooperative unit was started in the year 1985 with an initial capital investment of Rs. 2 lakhs and also with running capital of Rs. 2 lakh. Initial capital was private while running capital was obtained as loan from Khadi board. yearly sale proceeds approxmate. Rs. 2.5-3 lakhs, while net profit per year approximates Rs. 25,000/-

ROLES

The major role of this organisation is production. Training is also imparted but in a very restricted manner. The processed products are pickles, syrups and murabba. Pickles are prepared from amla, lemon, chillies, ginger, garlic, turnip and cauliflower. Murabba is prepared from amla, apple, carrot, ber etc. Yearly total production of pickle is 6 tonnes and that of murabba is also 6 tonnes. The production of syrup is 200 ltrs.

Training of one week duration has been imparted to 5 persons including one male and 4 females.

Organisational Resources : The plinth area of the building is 200 sq. yds,out of which production area is 180 sq. yds. The organisation possesses machines and equipments worth Rs. 30,000/-. The machine and equipment were purchased locally and from Delhi.

Organisation is a cooperative society consisting of 16 members. The actual functionaries include one supervisor (graduate) and eight labourers.

The supervisor looks after the complete management of the processing unit and the labourers carry out physical works.

Input Supply : Raw material is purchased from the local market and Delhi.

Marketing : Marketing of the products is done locally throughout the state through retailers and wholesalers.

Linkages : There are no wide linkages except for loan from Khadi board and with retailers for whole sale purposes.

Personal, social and psychological profile of rural families: The data in Table 5.6 depict that most of respondents (52.30%) and (53.85%) were of old age belonging to nuclear family and majority of them had medium sized family. It was further observed that majorityof the respondents (56.92%) were of middle caste and family education status.

Majortity of the women (64.62%) were illiterate, whereas 55.39 per cent men were middle pass. Majority of them (52.30%) had status of mother-in-law and father-in-law in the family. As regards their social participation, 50.77 per cent women were not the members of any organisation, however, 55.38 per cent men were members of one organisation. It was further observed that 92.31 per cent of the men were having frequent urban contact, whereas 50.77 per cent of women were having seldom urban contact. As regards their extension contact, majority of the respondents (72.31%) had no contact with extension personnel while majority of the women respondents (69.23%) were somewhat economically motivated and 44.62 per cent of them were less risk-oriented. Most of the women respondents (47.69%) were having moderately favourable attitude whereas 43.08 per cent men had highly favourable attitude and majority (61.54%) had low level of knowledge. The data in Table 5.7 depict the majority of the families (64.62%) were pursuing fruit and vegetable procesing at commercial level and 55.38 per cent had medium size of machinery and equipment possession. As regards training, 61.54 per cent of the men had technical training whereas none of the women respondents had obtained any training. It was further observed that 46.16 and 69.23 per cent of the families had easy availability of raw material and machinery and equipment possession, respectively.

Major processing technologies being practised by rural families

The rural families practised various technologies for the processing of fruits and vegetables. The results have been presented hereafter:

Table 5.6. Personal, social and psychological profile of rural families

Sr. No.	Attributes	Categories	Women n=65	Men n=65
		PERSONAL ATTRIBUTES		
1.	Age	Young (below 25 yrs)	-	-
		Middle (25-50 yrs)	31 (47.70)	30 (46.15)
		Old (Above 50 yrs)	34 (52.30)	35 (53.85)
2.	Education of Respondents	Illiterate	42 (64.62)	-
		Primary	20 (30.77)	5 (7.69)
		Middle	3 (4.61)	36 (55.39)
		High School	-	24 (36.92)
		Technical education		
		Graduate		
		Post graduate		
3.	Family size	Small (Upto 5 members)	-	
		Medium (5-10 members)	39 (60.0)	
		Large (Above 10 members)	26 (40.0)	
4.	Family education status	Low (0-1)	25 (38.46)	
		Medium (2-3)	36 (55.38)	
		High (3-4)	04 (6.16)	
5.	Status in Family	Daughter	-	
		Daughter-in-law	31 (47.70)	
		Mother-in-law	34 (52.30)	
		Son (Unmarried)		
		Husband		31 (47.70)
		Father-in-law		34 (52.30)

(Contd.)

Sr. No.	Attributes	Categories	Women n=65	Men n=65
SOCIAL ATTRIBUTES				
1.	Caste	Lower	-	
		Middle	37 (56.92)	
		Higher	28 (43.08)	
2.	Family type	Nuclear	34 (52.30)	
		Joint	31 (47.70)	
3.	Urban contact	Never	-	-
		Seldom (between	33 (50.77)	
		Frequently	6 months	5 (7.69)
			32 (49.23)	60 (92.31)
4.	Social participation	Nil	33 (50.77)	25 (38.46)
		Member of one organisation	32 (49.23)	36 (55.38)
		Member of more than one organisation	-	4 (6.16)
		Office holder	-	
5.	Extension contact	Frequnt (3-4 times)	-	
		Seldom (2-3 times)	-	
			18 (27.69)	18 (27.69)
		Never	47 (72.31)	47 (72.31)
PSYCHOLOGICAL ATTRIBUTES				
1.	Attitude	Less favourable (10-22)	16 (24.62)	7 (10.77)
		Moderately Favourable (23-36)	31 (47.69)	30 (46.15)
		Favourable (37-50)	18 (27.69)	28 (43.08)
2.	Economic Motivation	Less motivated (0-3)	20(30.77)	20 (30.77)
		Somewhat motivated (3-6)	45 (69.23)	23 (35.38)
		Highly		22 (33.85)

(Contd.)

Sr. No.	Attributes	Categories	Women n=65	Men n=65
		Motivated (6-9)		
3.	Risk Orientation	Low (6-15)	29 (44.62)	7 (10.77)
		Medium (15-25)	18 (27.69)	48 (73.85)
		High (25-35)	18 (27.69)	10 (15.38)
4.	Knowledge	Low	40 (61.54)	21 (32.31)
		Moderate	20 (30.77)	31 (47.69)
		High	5 (7.69)	13 (20.00)

(Figures in parentheses indicate percentages)

Table 5.7. Resource system profile of rural families.

Sr. No.	Resource system variables	Category	Women n=65	Men n=65
1.	Type of unit	Home scale	23 (35.38)	
		Commercial level	42 (64.62)	
2.	Machinery & equipment possession	Low possession	-	
		Medium Possession	36 (55.38)	
		High possession	29 (44.62)	
3.	Training	No training	65 (100.0)	25 (38.46)
		Technical training	-	40 (61.54)
4.	Availability of raw Material	Not available in time	10 (15.38)	
		Available at few places	25 (38.46)	
		Easily available	30 (46.16)	
5.	Availability of machinery and equipment	Not needed		
		Available at selected places	20 (30.77)	
		Easily available	45 (69.23)	

(Figures in parentheses indicate percentages)

Table 5.8. Extent of managerial, processing & marketing role performance by rural women in fruit and vegetable processing.

Sr. No.	Managerial roles	Extent of performance (n=65)			Total Score	Mean Score	Rank
		Always (2)	Some times (1)	Never (0)			
1.	Making plan of work	10 (15.39)	40 (61.54)	15 (23.07)	60	0.92	IV
2.	Establishing processing priorities	45 (69.23)	15 (23.07)	5 (0.70)	105	1.61	II
3.	Identifying need for inputs such as raw material	58 (89.23)	7 (10.77)	0 (0.00)	123	1.89	I
4.	Arranging finances	-	15 (23.07)	50 (76.93)	15	0.23	VII
5.	Making contacts for finances	-	5 (7.70)	60 (92.30)	5	0.07	X
6.	Maintaining records for production, market price and sale	6 (9.23)	17 (26.15)	42 (64.62)	29	0.44	V
7.	Gathering the information about loaning schemes	-	20 (30.77)	45 (69.23)	20	0.30	VII

(Contd.)

Sr. No.	Managerial roles	Extent of performance (n=65)			Total Score	Mean Score	Rank
		Always (2)	Some times (1)	Never (0)			
8.	Procuring sound information from experimental stations	3 (4.62)	15 (23.07)	47 (72.31)	21	0.32	VI
9.	Modifying plans based on conditions.	30 (46.16)	20 (30.77)	15 (23.07)	80	1.23	III
10.	Making purchases of raw materials	-	10 (15.39)	55 (84.61)	10	0.16	IX
PROCESSING ROLES							
1.	Selecting	50 (76.93)	15 (23.07)	-	115	1.76	II
2.	Washing	60 (92.30)	5 (7.70)	-	125	1.92	I
3.	Chopping	6 (9.23)	50 (76.93)	9 (13.84)	62	0.95	IV
4.	Preserving	20 (30.77)	15 (23.07)	30 (46.16)	55	0.84	V
5.	Blanching	35 (53.86)	15 (23.07)	15 (23.07)	85	1.30	III
6.	Packaging	15 (23.07)	20 (30.77)	30 (46.16)	50	0.76	VI

(Contd.)

Sr. No.	Managerial roles	Extent of performance (n=65)			Total Score	Mean Score	Rank
		Always (2)	Some times (1)	Never (0)			
MARKETING ROLES							
1.	Identifying segments/targets	-	30 (46.16)	35 (53.84)	30	0.46	III
2.	Analysing competition	-	21 (32.30)	44 (67.69)	21	0.32	IV
3.	Planning seasonal marketing	-	38 (58.46)	27 (41.54)	38	0.58	I
4.	Packaging according to market	-	18 (27.70)	47 (72.30)	18	0.27	VI
5.	Checking the price of products	-	35 (53.84)	30 (46.16)	35	0.53	II
6.	Deciding for promotion	-	13 (20.0)	52 (80.0)	13	0.20	VII
7.	Making decision for distribution	-	20 (30.77)	45 (69.23)	20	0.30	V

(Figures in parentheses indicate percentages)

Table 5.9. Managerial, processing and marketing role performance by rural men in fruit & vegetable processing

Sr. No.	Managerial roles	Extent of performance (n=65)			Total Score	Mean Score	Rank
		Always (2)	Some times (1)	Never (0)			
1.	Making plan of work	30 (46.16)	25 (38.45)	10 (15.39)	85	1.30	VII
2.	Establishing processing priorities	20 (30.77)	20 (30.77)	25 (38.46)	60	0.92	VIII
3.	Identifying need for inputs	7 (10.77)	45 (69.23)	13 (20.0)	59	0.90	IX
4.	Arranging finances	60 (92.30)	5 (7.70)	-	125	1.92	II
5.	Making contact for finances	65 (100.0)	-	-	130	2.00	I
6.	Maintaining records for production, Market prices and sale	42 (64.62)	17 (26.15)	6 (9.23)	101	1.55	VI
7.	Gathering information about loaning	45 (69.23)	20 (30.77)	-	107	1.64	IV
8.	Procuring sound information from experimental stations	47 (72.31)	10 (15.39)	8 (12.30)	104	1.60	V

(Contd.)

| Sr. No. | Managerial roles | Extent of performance (n=65) | | | Total Score | Mean Score | Rank |
		Always (2)	Some times (1)	Never (0)			
9.	Modifying plans based on Conditions	15 (23.07)	30 (46.16)	20 (30.77)	60	0.92	VIII
10.	Purchasing raw material	58 (89.23)	7 (10.77)	-	123	1.89	III
PROCESSING ROLES							
1.	Selection	5 (7.69)	10 (15.38)	50 (76.93)	20	0.30	VI
2.	Washing	12 (18.46)	33 (50.77)	20 (30.77)	57	0.87	IV
3.	Chopping	25 (38.46)	34 (52.31)	6 (9.23)	84	1.29	III
4.	Blanching	15 (23.08)	15 (23.08)	35 (53.84)	45	0.69	V
5.	Preserving	35 (53.84)	15 (23.08)	15 (23.08)	85	1.30	II
6.	Packaging	50 (76.93)	15 (23.07)	-	115	1.76	I

(Contd.)

Sr. No.	Marketing roles	Extent of performance (n=65)			Total Score	Mean Score	Rank
		Always (2)	Some times (1)	Never (0)			
1.	Identifying segments	35 (53.84)	25 (38.46)	5 (7.70)	95	1.46	V
2.	Analysing competition	27 (41.53)	15 (23.08)	23 (35.39)	69	1.06	VII
3.	Plan marketing according to season	47 (72.31)	10 (15.38)	8 (12.31)	104	1.60	III
4.	Packaging according to market	30 (46.15)	20 (30.77)	15 (23.08)	80	1.23	VI
5.	Checking the price of products	45 (69.22)	15 (23.08)	5 (7.70)	105	1.61	II
6.	Deciding for promotion	52 (80.0)	5 (7.70)	8 (12.30)	109	1.67	I
7.	Taking the decision for distribution	40 (61.54)	18 (27.69)	7 (10.77)	98	1.50	IV

(Figures in parentheses indicate percentages)

The data in Fig. 5.4 show that all (100%) of the rural families were practising the technology of drying and use of salt with oil. Most of the families (53.86%) were making the products with the use of sugar.

Extent of performance of rural women in managerial, processing and marketing roles : The data on extent of performance of rural women in managerial, processing and marketing roles related to fruit and vegetable processing, have been presented in Table 5.8. It is revealed that maximum performance of women was for identifying need for inputs (I) followed by establishing processing priorities (II), modifying plans acording to conditions (III), making plan of work (IV) and maintaining records for production and sale (V). The performance of women was low in activities like making contacts for finances   (X). purchase of raw material (IX) arrangement of finances (VII), securing information about loaning schemes (VII) and seeking latest information from research stations. (VI). Women's low performance in arrangment of funds, purchases of raw material and seeking information from research stations might be due to the nature of duties, i.e., outside the home surroundings, which were basically acepted as man-oriented.

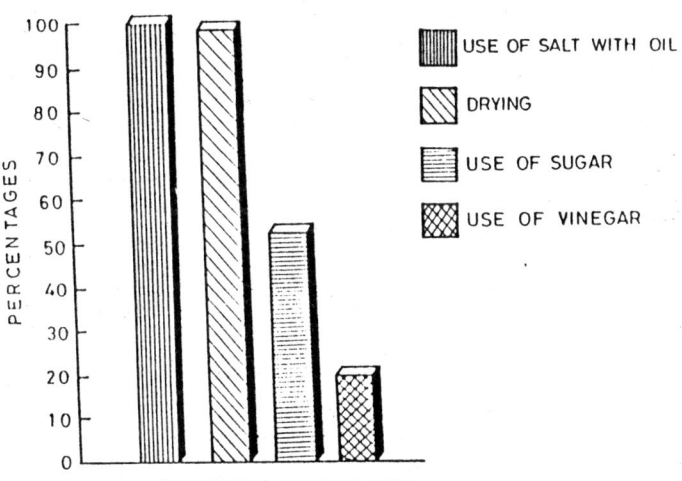

Fig. 5.4 : Processing technologies practised by rural families

It further shows that maximum performance of women in processing activities was in washing (I), selecting (II) and blanching (III). The performance of women was low in activities like packaeing (VI), preserving (V) and chopping (IV). Women's maximum performance in washing, selecting and blanching might be due to the fact that women were not trained for skilled activities. They were only performing the jobs that were not skill-oriented.

With regard to marketing roles, maximum performance of women was in planning according to season (I) followed by checking the price of products (II) and identification of segments (III).

Extent of performance of rural men in managerial, processing and marketing roles : The data on extent of performance of rural men in managerial, processing and marketing roles related to fruit and vegetable processing have been presented in Table 5.9. It is *revealed that maximum performance of men in managerial activities was in making* contact for finances (I), arrangement of finances (II), purchasing raw material (III), seeking information about loaning schemes (IV) and from experimental stations (V). The minimum performance of men was in activities like identifying need for inputs (IX), establishing processing priorities and modifying plans according to conditions (VIII), making plan of work (VII) and maintaining records for production, market prices and sale (IV).

As regards the processing roles, maximum performance of men was in activities like packaging (I) followed by preserving (II) and chopping (III), whereas minimum performance was in the those activities in which women perform maximum i.e., selection (VI), blanching (V) and washing (IV).

It further shows that in marketing roles men were performing mainly for the promotion of products (I), followed by checking the price of products (II) and planning the marketing of *products according to season (III)*.

Percentage of rural men and women's role performance in processing activities : The data in Fig. 5.5 & 5.6 show the distribution of men and women performance for processing activities on the basis of their percentages. The findings reveal that 55.38 per cent of women perform poorly in the managerial roles relating to fruit and vegetable processing, whereas 56.93 per cent of men had good performance.

For processing roles, 84.61 per cent of women had an average performance, whereas only 15.39 per cent of men and women had good performance. As regards the marketing roles, 73.39 per cent of men and women had good performance. As regards the marketing roles, 73.85 and 26.15 per cent of women had poor and average performance, respectively, in comparison to men performing good in marketing.

Data in Table 5.10 indicate that 't' test shows significant difference between the performance of men and women in managerial, processing and marketing roles of fruit and vegetable processing (t=33.90*, 6.47* and 42.55*, respectively) at commercial and home scale level.

Table 5.10. Difference between the performance of rural men and women in fruit and vegetable processing roles.

Sr. No.	Roles	Women n=65_ Mean (X)	Men n=65_ Mean (X)	't' values
1.	Managerial Roles	8.73	16.47	33.90*
2.	Processing Roles	48.09	38.92	6.47*
3.	Marketing Roles	6.86	13.09	42.55*

't' tab = 1.98 at 0.05 level of significance with n-2 d.f.
* Significant at 0.05 level of probability with 128 d.f.

Constraints faced by rural families in performing managerial, processing and marketing roles : The Fig. 5.7 presents the various constraints faced by rural families in performing roles in fruit and vegetable processing. It was observed from the findings that most of the respondents

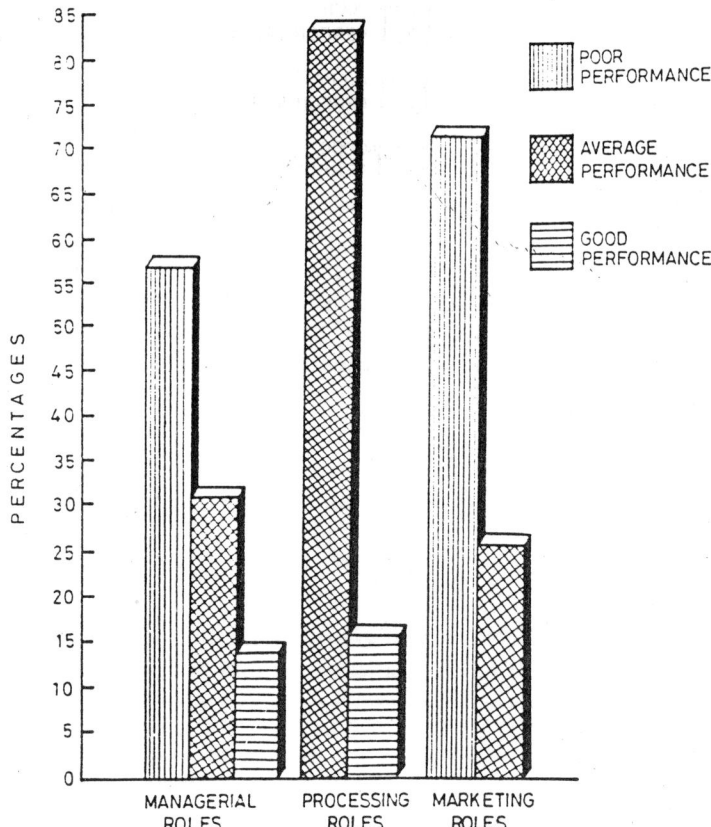

Fig. 5.5 : Percentage of rural women's performance in fruit and vegetable processing roles.

(89%) reported non-availability of modern processing technologies and lack of information about improved technologies (80%). A sizeable number of rural families (78%) mentioned lack of sufficient capital, lack of information about loaning schemes (60.64%), excessive burden of work and responsibility (56%), lack of recognition and appreciation in the family (53%).

Some of the other constraints indicated by the study were high cost and distant place for the availability of raw material (42%), difficulty in getting money from buyer after sale (36%), and availability of machinery and equipment at a high cost (33%).

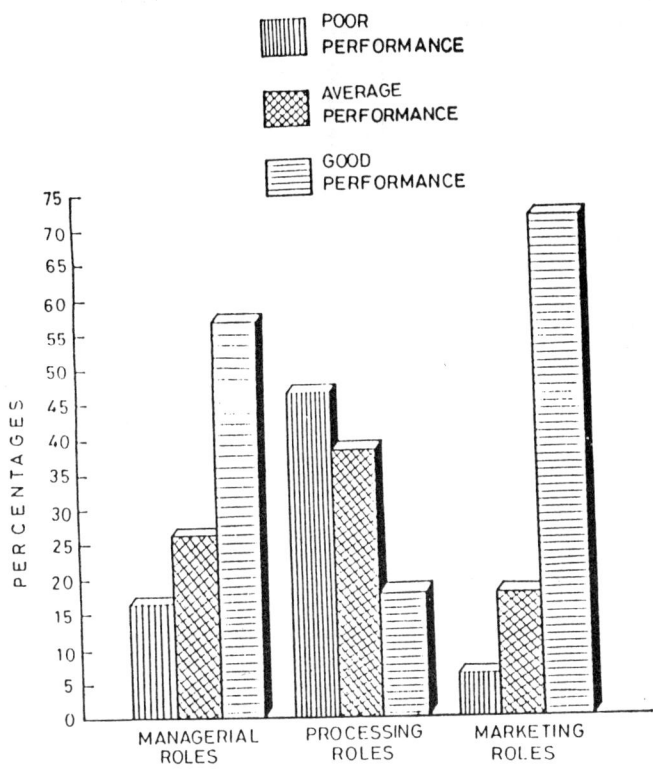

Fig. 5.6 : Percentage of rural men's performance in fruit and vegetable processing roles.

Data in Table 5.11 depict that rural families had somewhat linkages with rural families of their own village and very low linkage with families of other villages, scientists and field functionaries. No linkage was reported by rural families with input agencies and cooperatives. However, they had high linkages with traders for the purpose of selling their products.

The data in Table 5.12 show that rural families formed linkages with families of their own village through panchayat meetings, personal visit and fruit shows to some extent.

Panchayat meeting, personal visit and fruit shows were used to negligible extent to form linkage with rural families of other villages.

The data further show that personal visit, training and fruit shows were used to negligible extent to form linkage with scientists and field functionaries. However, none of the modes was used to have linkage with banks.

Table 5.11. Extent of inter-organisational linkages of institutions

Insti-tution	Fami-lies of own village	Fami-lies of other village	Scien-tists	Field funct-ion aries	Coop-erat-ives	Banks	Traders
Rural/ families	1.0	0.5	0.2	0.2	-	-	2.0

2 - High   1.9 - Somewhat   less than 1-Low

Table 5.12. Extent of mode used by institutions to form linkages with other organisations and institutions

Organisation/ Institution	Pan-chayat meeting	Coope-rative	Perso-nal	Train-ing	Shows
Rural families (own village)	0.5	-	0.6	-	0.2
Rural families (other villages)	0.4	-	0.3	-	0.2
Scientists	-	-	0.2	0.4	0.1
Field functionaries	-	-	0.4	0.4	0.2
Banks	-	-	-	-	-
Traders	-	-	1.0	-	-

* 1 and above - Mostly; 0.5 and 1 - Sometime; less than 0.5 - Negligible

Personal visits were mostly used by the rural families to communicate with traders.

Identification of appropriate technology of fruit and vegetable processing for disseminationin rural areas : Several technologies of fruits and vegetables have been identified by various researches. While disseminating a technology in the rural situation appropriateness of the technology should be assessed. The technology to be

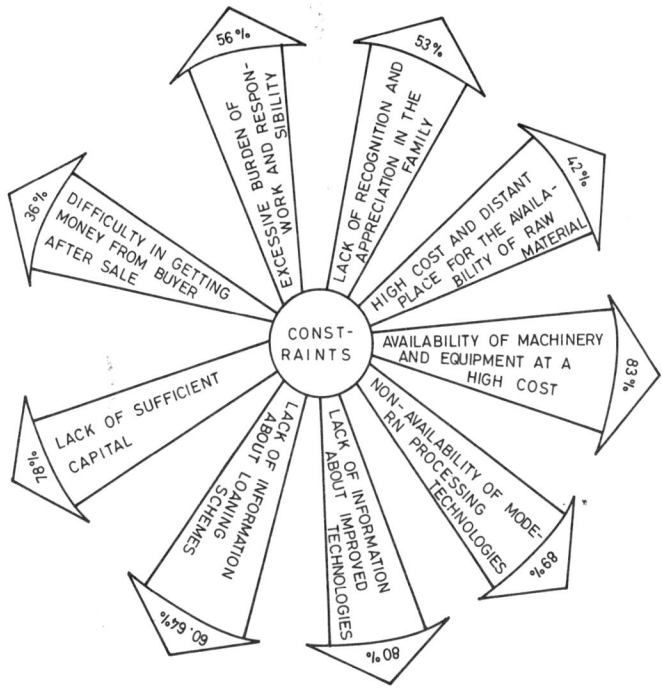

Fig. 5.7 : Constraints faced by rural families during processing of fruits and vegetables.

appropriate should meet the criteria of low cost, low input, low risk, rural bias, suitable for *small scale* application, use of local inputs and compatible with man's need for creativity. Therefore, appropriateness of the technology in the present context has been defined as one with low cost, low input, low risk type, rural bias, suitable for small scale application, use of local inputs and compatiable with man's need for creativity.

Identification of appropriate technology of fruit and vegetable processing : The appropriate assessment of the available technologies of fruit and vegetable (post-harvest,

Table 5.13. Identification of appropriate post - harvest technology

Technologies (post-harvesting stage)	Extent of appropriateness n=30			Total Weighted score	Ranks
	Most Appropriate (3)	Appropriate (2)	Least appropriate (1)		
Wax coating	8 (26.67)	15 (50.0)	7 (23.33)	61	IV
Evaporative cooling system/ zero energy chamber	20 (66.67)	10 (33.33)	–	80	I
Oiling	10 (33.33)	15 (50.0)	5 (16.67)	65	III
Disinfection by chemicals	5 (16.67)	10 (33.33)	15 (50.0)	40	V
Modern-cooling	15 (50.0)	10 (33.33)	5 (16.67)	70	II
Colouring	–	–	(30 (100.0)	30	VII
Sheetal pot	–	5 (16.67)	25 (83.33)	35	VI
Bamboo iceless refrigerator	–	5 (16.67)	25 (83.33)	35	VI

(Figures in parentheses indicate percentages)

Table 5.14. Identification of appropriate processing technology.

Technologies (Processing stage)	Extent of appropriateness n=30			Total Weighted score	Ranks
	Most Appropriate (3)	Appropriate (2)	Least appropriate (1)		
Fermentation	10 (33.33)	10 (33.33)	10 (33.33)	60	V
Sugar	20 (66.67)	10 (33.33)	–	80	III
Salt	25 (23.33)	5 (16.67)	–	85	II
Vinegar	15 (50.00)	10 (33.33)	5 (16.67)	70	IV
Drying (Sun-drying & solar drying)	30 (100.0)	–	–	90	I
Chemicals	5 (16.67)	10 (33.33)	15 (50.0)	50	VI

(Figures in parentheses indicate percentages)

Table 5.15. Identification of appropriate packaging technology

Technologies (Packing stage)	Extent of appropriateness n=30			Total Weighted score	Ranks
	Most Appropriate (3)	Appropriate (2)	Least appropriate (1)		
Ventilated polythene bags	20 (66.66)	5 (16.67)	5 (16.67)	75	II
Indigenous tin coating	10 (33.33)	5 (16.67)	15 (50.00)	55	III
Tray packing	25 (83.33)	5 (16.67)	5 (16.67)	85	I

(Figures in parentheses indicate percentages)

processing and packaging) was done through the experts of Horticultural Sciences and Foods and Nutrition on a standardized index. The results have been presented in Tables 5.13, 5.14 and 5.15.

On the basis of weighted scores, total ranks were assigned to different technologies (Table 5.13). The technology which got the first rank was considered as an appropriate post-harvest technology of fruit and vegetable for dissemination in rural areas. Evaporative cooling system/ zero energy cool chamber was judged as an appropriate post-harvest technology. Roy and Khurdiya (1986), Habibunnisa et al. (1988), Manick and Manimegali (1988), Sethi and Maini (1989), Kumari Sarala (1990) also recommended the use of evaporative cooling system/zero energy cool chamber to preserve the fruits and vegetables, to extend the storage life and to reduce the weight loss in storage.

Similarly, on the basis of weighted scores, drying ranked first as an appropriate technology for processing of fruits and vegetables with the help of solar driers. Similar findings were reported by Dhabade and Khedkar (1980); Harigopal and Tonapi (1980); Kalra and Bhardwaj (1981); Khurdiya and Roy (1986); Powar et al. (1988); Chaturvedi (1990); Kumari Sarala (1990); Jayaraman et al. (1991).

For packaging, tray packing ranked first as an appropriate technology for fruit and vegetables. Similar findings were reported by Maini et al. (1984).

Identification of appropriate technology on the basis of reviewed data : Review of the last one decade studies based on appropriate fruit and vegetable processing technologies for rural areas was made. The review consists of all the three stages, i.e., post-harvest, processing and packaging stage. The data are presented in Table 5.16, 5.17 and 5.18.

The data in Table 5.16 depict the appropriate technologies for the post-harvest stage. Data reveal that zero energy cool chamber was considered most appropriate because 37.94 per cent studies were conducted to test its appropriateness.

Similarly, the data in Table 5.17 reveal that for the processing, drying (77.27) is considered as the most appropriate for rural areas.

Table 5.16. Appropriate post-harvest technologies for rural areas.

Technologies		Frequency	(n=29)
1.	Wax coating	5	(17.24*)
2.	Zero energy cool chamber (Cooling)	9	(31.04)
3.	Oiling	11	(37.94)
4.	Disinfection	3	(10.34)
5.	Colouring	1	(3.44)

* Figures in parentheses indicate percentages.

Table 5.17. Appropriate processing technologies for rural areas.

Sr. No.	Processing technologies	Frequency	(n=22)
1.	Drying (Solar drying)	17	(77.27*)
2.	Use of salt and vinegar	4	(18.18)
3.	Use of sugar	2	(9.09)
4.	Fermentation	1	(4.45)

* Figures in parentheses represent percentages

Table 5.18. Appropriate packaging technologies for rural areas.

Packaging technologies	Frequency	(n=3)
Ventilated polythene bags	1	(33.33)
Indigenous tin coating	1	(33.33)
Tray packing	1	(33.33)

As it is clear from the data in Table 5.18, during the last one decade a very less number of studies was conducted on the appropriateness of the technology for packaging stage. On the basis of judges consensus and past researches conducted druing the last decade in the field of appropriate processing technology for rural areas, solar drying emerged as the best appropriate processing technology. A comparative performance testing of natural convection solar dryer with chimney and open sun-drying was conducted by drying three vegetables, i.e., chillies. The natural convection solar dryer (Chimney) designed and fabricated by the college of agricultural engineering (Fig.15).

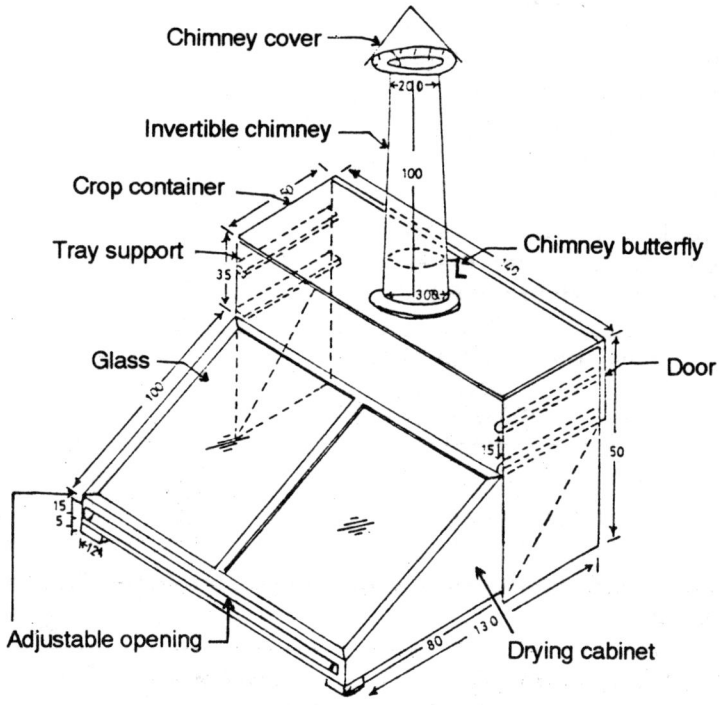

Fig 5.8 : Cabinet type natural convection solar dryer

Method

Green peas : Fresh peas were purchased from the local market. Green peas after depoding were washed in running water and then blanched in boiling water containing 0.5% potassium metabisulphite (KMS) for 3-4 minutes.

Potato chips : Potatoes were washed and peeled manually. One cm thick slices were prepared with hand slicer and blanched in boiling water containing 0.5% KMS solution for 3-4 minutes.

Chillies (green) : Full grown pods of bright green colour were purchased from the market. Pods were slit and seeds were removed. No treatment was given to the chillies.

All the vegetables treated as above were dried in a natural convection solar dryer (chimney) and in direct sun drying. Three kg of chillies were kept separately in each condition. The experiments were continued till the produce achieved its equilibrium moisture content. Hourly inside temperature was measured while carrying out the drying trial from 9.30 AM to 4.30 PM. Daily loss in moisture was measured by weighing the produce in the evening, contents of the product are presented in Table 5.19.

Table 5.19. Comparative moisture content in natural convection solar dryer with chimney and open sun-drying (initial m.c. 80%)

	Moisture content of dried product at the end of day	
Days of Exposure	**Natural convection Solar dryer with chimney**	**Open sun Drying**
1.	66.0	79.0
2.	57.0	72.0
3.	16.0	71.0
4.	6.0	69.0
5.		67.0
6.		61.5
7.		59.5
8.		55.5
9.		49.5
10.		46.5
11.		39.0
12.		37.5
13.		32.0
14.		27.5
15.		23.5
16.		19.0
17.		15.0
18.		10.5
19.		8.5
20.		6.5

Moisture content of chillies was reduced from 80% to 6% in four days in natural convection solar dryer with chimney and in 20 days in open sun-drying. Quality of drying and stress concentration at the surface of the chilly was better in natural convection dryer because the produce was not exposed to the sun shine and there was no forced convection of hot air. Colour of the dried chillies was brownish in natural convection solar dryer (Chimney) and it was blackish in open sun-drying. At this moisture level, the quality of the vegetables does not deteriorate as reported by Kalra and Bhardwaj (1981) and Yadav and Malviya (1992).

Efforts were made to maintain the drying temperature in the range of $50 \pm 5^{\circ}C$. For this, the glass over the black body of the dryer was partially covered with a white plastic sheet, which reflected the sun shine and temperature reaching the safe limits.

Potato Chips

After the treatment the potato slices were dried with a tray load of 1000g/3000 sq.cm. in natural convection solar drier (chimney) and direct sun. The initial moisture content of potato slices was 75.83% and increased to 82.96% during blanching. The moisture content of potato chips was reduced to 4.4 and 6.0 per cent (Table 5.20) in the natural convection solar dryer with chimney and open sun drying, respectively. In case of dehydration the dryer took 3 hours for potato chips as compared to 4 hours needed in open sun drying.

Table 5.20. Solar dehydration test for potato chips.

Material	Natural convection Solar dryer (Chimney)		Direct sun drying	
	Moisture %	Drying period	Moisture %	Drying period
Potato chips	4.4	3hrs	6.0	4 hrs

The data are the average of 3 observations.

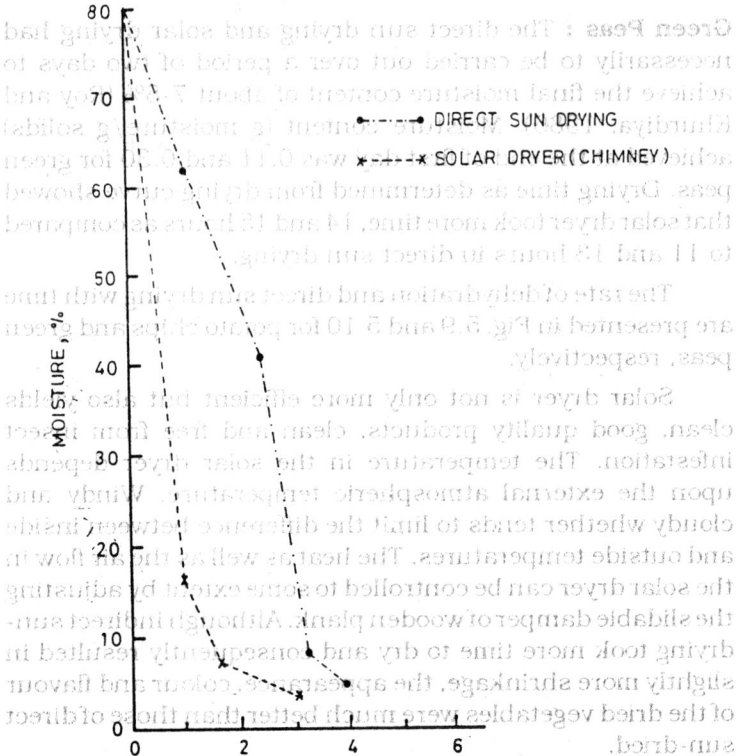

Fig. 5.9 : Dehydration pattern of potato chips in solar dryer.

that rural families having more education, status of housewife, having maximum extension contact, more of family urban contact, favourable attitude towards food processing, more knowledge about food processing and training tend to perform good management in fruit and vegetable processing. Further, negative association of constraints with performance of managerial roles suggests that more of the constraints are faced by rural families in managing fruit and vegetable processing, resulting in their decreased performance.

Correlation coefficients of processing roles of rural families in fruit and vegetable processing with their antecedent variables : The correlation coefficients presented in Table 5.22 show that nine variables, viz., education of respondent,

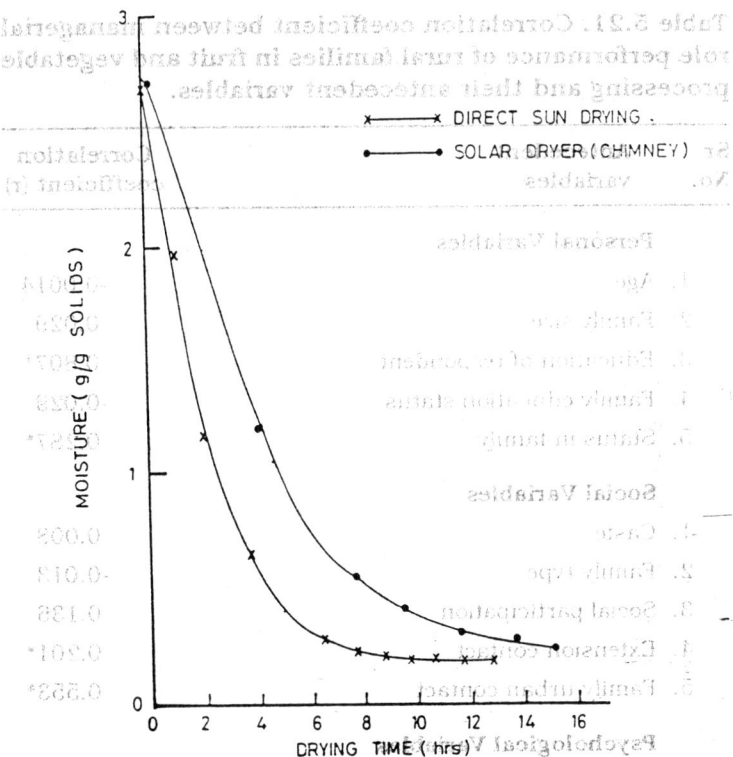

Fig. 5.10 : Drying Curves For Green Pea

extension contact, family urban contact, type of unit, knowledge, availability of raw material, training, machinery and equipment possession and constraints had significant association with role performance of rural families. However, it was negative for five variables, i.e., education of respondent, extension contact, family urban contact, training constraints and positive for others. It may be inferred that those who had more knowledge about fruit and vegetable processing, commercial type of unit, easy availability of raw material and more of machinery and equipment tended to perform better in the processing of fruits and vegetables. These findings are in tune with the previous observations of Kherdy and Sahay (1972); Kadam and Valunj (1982); Sohi and Kherde (1985); Sharma *et al.* (1988).

Further, negative association of education of respondent, extension contact, family urban contact, training and

constraints with rural families, processing roles allude that less education of respondent, less extension, family urban contact and lack of training decreased their performance.

Correlation coefficient of marketing roles of rural families in fruit and vegetable processing with their antecedent variables : The results of correlational analysis presented in Table 5.23 show that seven variables were significantly correlated. But six of them, viz. education of respondent, status in family, extension contact, family urban contact, knowledge and training had positively significant association with marketing roles of rural families and the seventh, i.e., constraints faced by rural families were negatively associated with marketing roles of rural families. It may be inferred that respondents, who were better qualified, belonged to a status of mother-in-law, had more of extension and family urban contact, more knowledge about fruit and vegetable processing and training tended to perform better in fruit, and vegetable processing. It stands justifiable and reasonable also and lends support to earlier observation of Kherde and Sahay (1972); Kadam and Valunj (1982); Sohi and Kherdy (1985); Sharma *et al.* (1988).

Further, negative associationof constraints faced by rural families with marketing roles suggested that more of the constraints faced by rural families decreased their performance in fruit and vegetable processing.

Correlation coefficients of overall role performance of rural families infruit and vegetable processing with their antecedent variables : The correlation coefficients presented in Table 5.24 reveal that role performance of rural families was significantly and positively associated with seven antecedent variables, viz., education, caste, family urban contact, availability of raw material, training, type of unit, machinery and equipment possession and negatively associated with one variable, i.e., constraints faced by rural families. It may be inferred that those rural families who were better qualified, belonged to higher caste, had more of family urban contact, easy availability of raw material, training and more of machinery and equipment, tended to perform better roles in fruit and vegetable processing. The present

Table 5.23. Correlation co-efficient between role performance (marketing role) of rural families in fruit and vegetable processing and their antecedent variables.

Sr. No.	Antecedent variables	Correlation coefficient
	Personal Variables	
1.	Age	0.012
2.	Family size	-0.047
3.	Education of respondent	0.803*
4.	Family education status	-0.018
5.	Status in family	0.285*
	Social Variables	
1.	Caste	0.011
2.	Family type	0.031
3.	Social participation	0.099
4.	Extension contact	0.278*
5.	Family urban contact	0.583*
	Psychological Variables	
1.	Attitude	0.168
2.	Economic motivation	0.061
3.	Risk orientation	0.123
4.	Knowledge	0.278*
	Other Variables	
1.	Type of unit	0.020
2.	Availability of raw material	0.033
3.	Availability of machinery & equipment	0.000
4.	Training	0.834*
5.	Machinery & equipment possession	0.043
6.	Constraints	-0.768*

* Significant at 0.05 level with 128 d.f.

families in processing roles with antecedent variables have been presented in Table 5.27. An examination of the data show 91 per cent variation in processing roles of rural families. This could be explained by nine antecedent variables, viz., education of respondent, extension contact, family urban contact, type of unit, availability of raw material, training, machinery and equipment possession, knowledge and constraints. Individually, only one variable, i.e., type of unit had significant positive regression coefficients with processing roles of rural families. Therefore, it can he implied that this variable contributed to a large extent in influencing processing roles of rural families.

Regression coefficients of marketing performance of rural families with antecedent variables : The regression coefficients presented in Table 5.28 reveal that all the nine antecedent variables, viz., education of respondent, extension contact, status in family, family urban contact, training, attitude, knowledge and constraints faced by rural families together contributed 85.1 per cent variation in marketing roles of rural families. It is also observed that two variables, viz., family urban contact and training exhibited significant regression coefficients towards role performance. Thus, it implies that these two variables were important predictors of performance of rural families in marketing roles.

Regression coefficients of rural families total role performance in fruit and vegetable processing with antecedent variables : The regression Co-efficients of role performance of rural families in fruit and vegetable processing with antecedent variables have been presented in Table 5.29. An examination of Table 5.29 shows 88.51 per cent variation in performance of roles in fruit and vegetable processing by rural families. This could be explained by eight antecedent variables, viz., education of respondent, caste, family urban contact, type of unit, availability of raw material, training, machinery and equipment possession and constraints faced by rural families. Individually, only one variable, i.e., type of unit had significant positive regression coefficient in role performance of fruit and vegetable processing. Therefore, it can be implied that this variable contributed to a large extent

Table 5.25. Correlation co-efficient between linkages of rural families in fruit and vegetable processing and their antecedent variables.

Sr. No.	Antecedent variables	Correlation coefficient
	Personal Variables	
1.	Age	0.047
2.	Family size	-0.128
3.	Education of respondent	0.075
4.	Family education status	-0.052
5.	Status in family	-0.004
	Social Variables	
1.	Caste	0.159
2.	Family type	-0.113
3.	Social participation	0.099
4.	Extension contact	-0.094
5.	Family urban contact	-0.007
	Psychological Variables	
1.	Attitude	0.035
2.	Economic motivation	0.172*
3.	Risk orientation	0.073
4.	Knowledge	0.044
	Other Variables	
1.	Type of unit	0.773*
2.	Availability of raw material	0.725*
3.	Availability of machinery & equipment	0.000
4.	Training	0.002
5.	Machinery & equipment possession	0.529*
6.	Constraints	-0.147

* Significant at 0.05 level with 128 d.f.

Table 5.29. Regression Co-efficients of rural families role performance in fruit and vegetable processing with Significant Independent Variables.

Sr. No.	Variables	Regression Co-efficient	't' values
1.	Education of respondent	0.87	1.19
2.	Caste	-0.20	-0.39
3.	Family urban contact	0.62	0.87
4.	Type of unit	29.58	11.21*
5.	Availability of raw material	-0.02	-0.04
6.	Training	0.41	0.70
7.	Machinery and equipment-possession	0.22	1.77
8.	Constraints	0.31	1.25

$R^2 = 0.90$

F value = 88.51*

* Significant

resources to accomplish a specified change in human behaviour. It looks into what is needed and then plan as to how to achieve. In the present framework it has been defined as the broadlines for promoting fruit and vegetable processing among rural areas by different sectors, i.e., governmental and non-governmental organisations to give effect to the greater participation of women in processing activities for the rural development. This strategy is based on the Triple 'A' (Fig. 5.11) approach, i.e., assessment, analysis and action.

Assessment : The cycle starts with the assessment of existing status for rural families, knowledge about different techniques and technologies about fruit and vegetable processing; their attitudes towards food processing as an economic activity; their extent of practising food processing at different levels and their linkages with other organisations and institutions working in this field.

Table 5.30. Regression Co-efficients of rural families Linkage in fruit and vegetable processing with Significant Independent Variables.

Sr. No.	Variables	Regression Co-efficient	't' values
1.	Type of unit	14.65	4.50*
2.	Availability of raw material	0.34	0.48
3.	Machinery & equipment-possession	0.08	0.54
4.	Economic motivation	0.50	1.29

$R^2 = 0.61$

F value - 38.78*

* Significant

Accordingly, there is a specific purpose of an assessment of the aspects involved in a particular rural family as mentioned here in connection with food processing at different levels. Therefore, a definite degree of frequency in undertaking assessment of all these aspects in a series is essential to achieve the objectives of initiating and extending food processing units and rendering them remunerative to make the family economy viable.

Analysis : Analysis refers to the critical evaluation of the existing status of knowledge, attitude, role performance and linkages.

Analysis of individual family attributes and sum total role performance facilitates pinpointing adequacy and inadequacy of the various contributory factors and also reveals the negative or positive contribution of linkages.

Action : The proposed strategy for implementation will be the action.

Decision making for an action requires complete knowledge of the attributes of a rural family, role performance and contribution of linkages. This requisite information is

men regarding fruit and vegetable processing technologies, financial and technical institutes (Fig. 5.12).

APPROPRIATE TECHNOLOGIES

Majority of rural families were practising the age-old technologies like sun drying and use of salt and oil in preserving fruits and vegetables at home scale. Based on the secondary data and from judges' consensus, the technology of drying with the help of solar dryer was found appropriate for processing fruits and vegetables.

ROLE PERFORMANCE

Majority of the respondents were processing fruits and vegetables at commercial level but the produce is cornered by the traders at a very low price to the producers.

Women's performance in managerial (55.38%) and marketing (73.85%) roles was very poor in comparison to 56.93 and 72.31 per cent men performing good in managerial and marketing. However, 84.61 per cent women were performing average in processing roles as compared to 38.46 per cent men performing average.

LINKAGES

None of the families had any linkage with financial institutions. However, 61.53 per cent men had linkages with field functionaries of community food and extension unit. Very less percentage of rural families had linkages with families of their own village and of other villages.

It is against this existing status of fruit and vegetable processing at institutional level that a strategy has been evolved which strives for the greater participation of rural women in food processing activities.

This developed strategy lays stress mainly on three components, i.e., What, Bywhom and How. For promoting women's participation in fruit and vegetable processing (Fig. 5.13). What means what is needed to promote women's participation in fruit and vegetable processing. By wModernns who should take the action. How means as to how to achieve what is needed in promoting participation of women in fruit and vegetable processing.

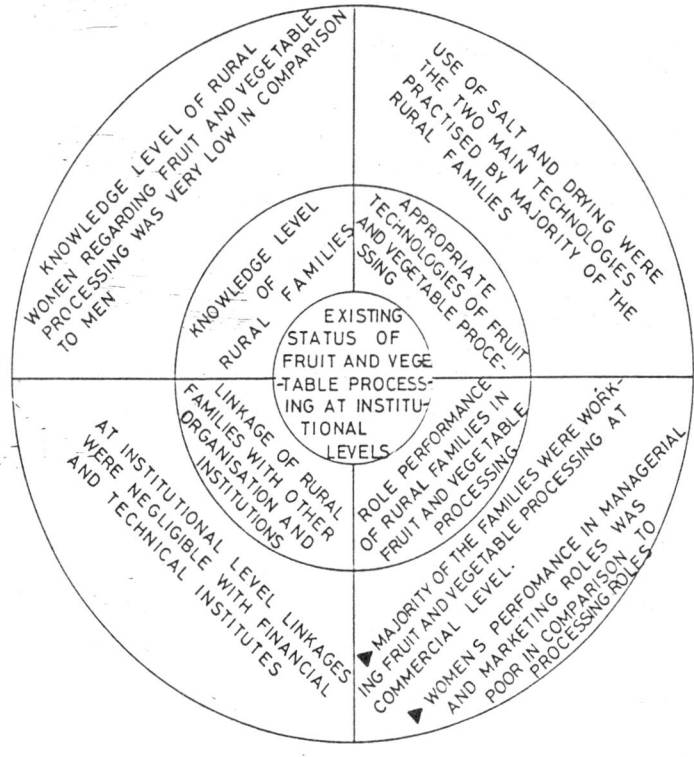

Fig. 5.12

Popularisation of proven identified appropriate technologies : The foremost action to be taken to promote participation of rural women in fruit and vegetable processing in rural areas is to popularise the identified appropriate technologies among them. Popularisation of technologies should be done by home science colleges by conducting food demonstration camps, developing material for mass media, conducting food demonstration camps, developing material for mass media, conducting training for rural youth both women and men. Home science faculties in agricultural universities can get support in education. research and extension in their endeavour to find appropriate processing technologies and in the transper of technology. Training should be organised in collaboration with rural development programmes, mahila mandals and voluntary organigations. Training should be held at colleges, krishi vigyan kendras and at village level depending on situations.

Promoting food processing as cottage unit at village level

Agro-industries have a good deal of potential in serving the rural economy. First, it helps in generating more employment opportunities for rural people. Agro industries in rural areas can also minimise the wastes in processing agricultural output. For instance, development of a large number of tiny rural industries in and around the place of production of crops such as tomato, potato, grapes, ber, guava etc., could help in minimising the seasonal losses. It will also check mobility of rural masses towards urban areas in search of employment. However, the mainpoint is that the growth of agro industries sector has not been on the line that is ideal for rural development. Today many agricultural resources are transported from rural areas to the urban industry centres for processing. Thus, the employment opportunities are lost in rural areas. The establishment of the Ministry of Food and Agro Industry at the centre is a major welcome change. Now the government should review the new agricultural and industrial policy to give greater emphasis to the employment oriented agro industries by providing financial assistance and formulating new policies so that these cottage units should receive a real fillip.

Women's Cooperatives. to meet the high cost of installation, a women's cooperative or a federation should be formed. These cooperative should develop a marketing network at village, state and national level to make the product available at the door step of the consumer. This could be achieved only if they are all brought under a centralized organisation so that they move from 'regional' to the 'national' level operation.

Self Employment : Employment opportunities offered by agro industry are plenty to the farm population and entrepreneurs seeking self-employment. Cottage scale units

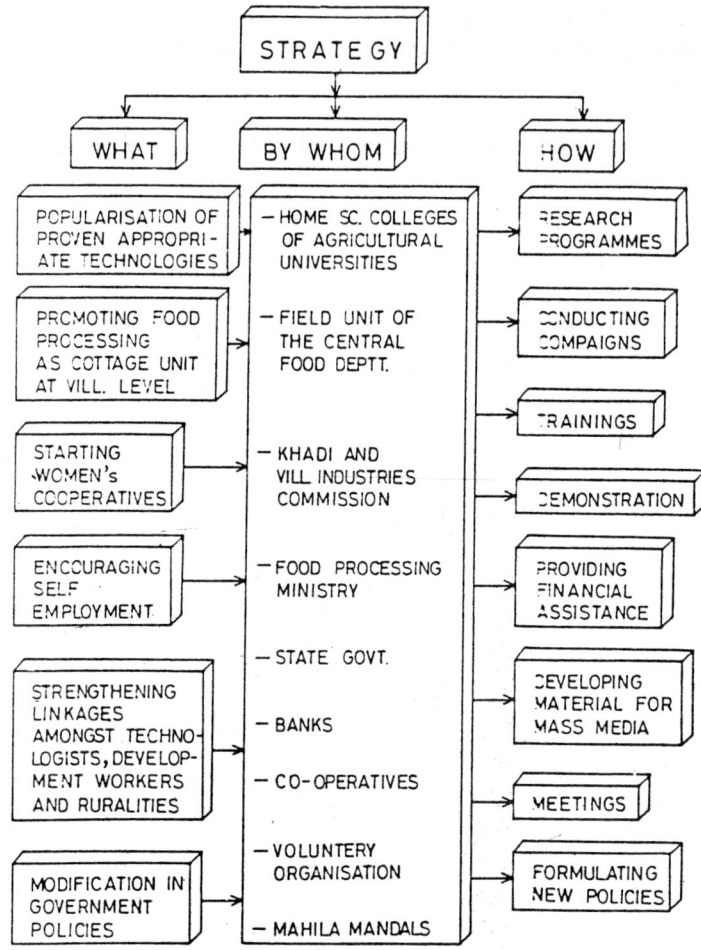

Fig. 5.13 : Strategy for promoting participation of women in fruit and vegetable processing at institutional level..

particularly offer self employment opportunities. Entrepreneurs with limited capital can set up processing units with limited overheads and marketing expenses in rural areas. Home science colleges should help the rural women in development of entrepreneurial competency in them by providing them training in related area and by motivating them. KVIC also implements its programme through state KVI Boards by extending needed assistance to individuals and co-operatives such as financial assistance, training support, raw material and marketing support if needed.

Strengthening linkages among technologists, development workers and ruralites

Research, teaching and extension are of prime importance in the growth and development of food processing industry in rural areas. New ideas and new basic knowledge can be developed and enhanced only through situational original research. the prevailing environments, ills, weaknesses, inadequacies in the rural areas cannot be overlooked for development of this industry. The socio-economic status of the people, their sources, resources, their ignorance, illiteracy, proverty, immobility etc. constitute the main impediments. The agricultural conditions, crop potentials etc. have to be planned to solve the problems. For this, concerned scientists are required to visit the rural areas very often for identification of problems and planning the desired research projects. Cost-effective technology is the real need of the ruralites. After development of such technologies their dissemination to the ruralites through development workers is very essential. If research findings do not reach the consumers then such researches become meaningless. Therefore, awareness of the ruralites as to what they can do for increasing their income has to be given top priority. It is only proper education through the development workers which will motivate the rural people to start agro based food processing units which can supplement their income and help in raising the living standard. Traditionally, women handle food and are familiar with the skills of food processing. In order to impart the latest knowledge and skills in rural areas, participation of rural women has to be increased as a special measure and rightly so. Low cost food processing technologies offer excellent opportunities for women in production of processed foods in rural areas. This can be achieved only through development of extension workers belonging to home science faculties of agricultural universities. The extension workers have to carry out two way traffic, i.e., they are to transmit technologies to the users very effectively and also to pass on the feed back on problems of users to the researchers. If this communication system is developed in this manner then there will be no gap in the to and fro communication.

Presently the village economy is being harnessed for satisfying urban needs, which ultimately results in the villages remaining in their backwardness and economic stagnation. If the development of agro based food processing units in rural areas is under-taken on a large scale and in as shorter time as possible, then the following changes can be achieved :

1. The women will be able to create economic viability in the household economy. They can certainly bring about dynamism in the stagnant and semi-stagnant rural economy.

2. Women can increase their earning capacity and thus spin income generating activities.

3. The status of socially backward and landless labour classes is possiblle only through providing non-farm employment at their door steps.

4. This will generate a sense of security and confidence amongst rural people for overcoming uncertainty in agricultural income and providing self employment to the landless labour.

5. The locally available untapped resources will be used effectively. The dormant ability and skills of the women will come into display through extension education programmes.

Modification in Government Policies : Agro-based industries have to play a vital role in the economic devlopment of the country. With a view to provide a needful thrust and to accelerate the growth, the Government should remove the taxes and duties especially those levied on tinplate and containers, glass bottles and closeres and excise on finished products for stabilizing prices at a reasonable and bearable level. There should also be a removal of the entry taxes and reduction of sales tax on processed foods to create an urban market for the rural product and thus generating a reverse flow of income from urban to rural areas. This would ultimately reduce rural migration into urban areas.

With a modified policy framework as mentioned above, rural India will acquire a place in sun for its fruits, vegetables and other agro products that would soon become enviable, not in the long future. Thus, government should recognise the rural food processing industry as a golden goose to be nurtured and not to be hacked.

For the development of food processing industry at different scales in rural areas, a liberal and concessional financing is required because the rural people are economically very backward. For initiation of the development, it is government initiative on all fronts which can motivate the ruralites, particularly women to start food processing units. It is definite that with financial provisions as suggested, processing industry can develop very extensively resulting in increasingly large production of the different food products. Obviously development of the food processing industry is dependent upon availability of liberal credit along with other inputs. But for maintaining the production momentum, development of viable and effective marketing system is another important aspect. Only a good marketing system can sustain the momentum of production. Production without the proper marketing system can collapse in no time which does not prove remunerative to the industry people, discourages them to keep up the efforts and ultimately compels them to close the production units. It is, therefore, obvious that marketing system is not to be lost sight of and it has to be developed along with production development if it is not considered to develop it before production initiative. It is suggested that marketing system can be viable and effective when it has a well knit network, i.e., three tier system, i.e., marketing within the state, interstate or country level marketing and international marketing. Further, it is suggested that marketing should be adopted with effective monitoring for the consumption areas, so that transport of products to different points of the sale is arranged and implemented without losing any time and without causing any inconvenience to the producers. It is also suggested that marketing system should become a belonging of the producers and rural people, particularly women be employed so that ownership feeling of the people sustains the marketing system with dynamism.

Strategy for promoting women's participation in fruit and vegetable processing organisation : This strategy for the organisations to promote women's participation in fruit and vegetable processing is also based on the Triple 'A' (Fig. 5.14) approach, i.e., assessment, analysis and action.

Assessment : The cycle starts with the assessment of appropriate fruit and vegetable processing technologies practised at various levels of organisations, i.e., large scale, small scale, training, khadi gramudyog and cooperative unit, their role performance and inter-organisational linkages.

There is always a specific purpose of assessment. Since this assessment pertains to food processing units of small, medium and large scale and also the training unit, the assessment of various facets of a particular organisation will ultimately reflect upon the quality and quantity of the products and the economics of their production. Hence assessment is often necessitated to bring about an overall improvement in an organisation.

Analysis : Analysis refers to the critical evaluation of the existing status of fruit and vegetable processing technologies at various levels of organisations, their performance in processing roles and inter-organisational linkages. Analysis of an organisation in terms of input and output will involve critical study of all the factors contributing to production efficiency. Until and unless it is done it is not possible to pinpoint the productive and unproductive factors involved in any of the food processing units and the training unit.

Action : The proposed strategy for implementation will be the action.

Decision for taking an action requires complete knowledge of all the production processes. For generation of the desired information, specific assessment and analysis are most essential. For improving an organisation like processing unit/ training unit, a definite degree of frequency for carrying out assessment, analysis and action in tandem is very essential.

Existing status of fruit and vegetable processing at various levels of organisation

Large scale units : The two large scale units were studied and study has brought to light that in Pan Food ltd. Various processing technologies are being practised such as use of sugar, use of salt, use of chemical and other preservatives, drying, canning and bottling, while Pachranga International is using only one technology, i.e., use of salt and oil.

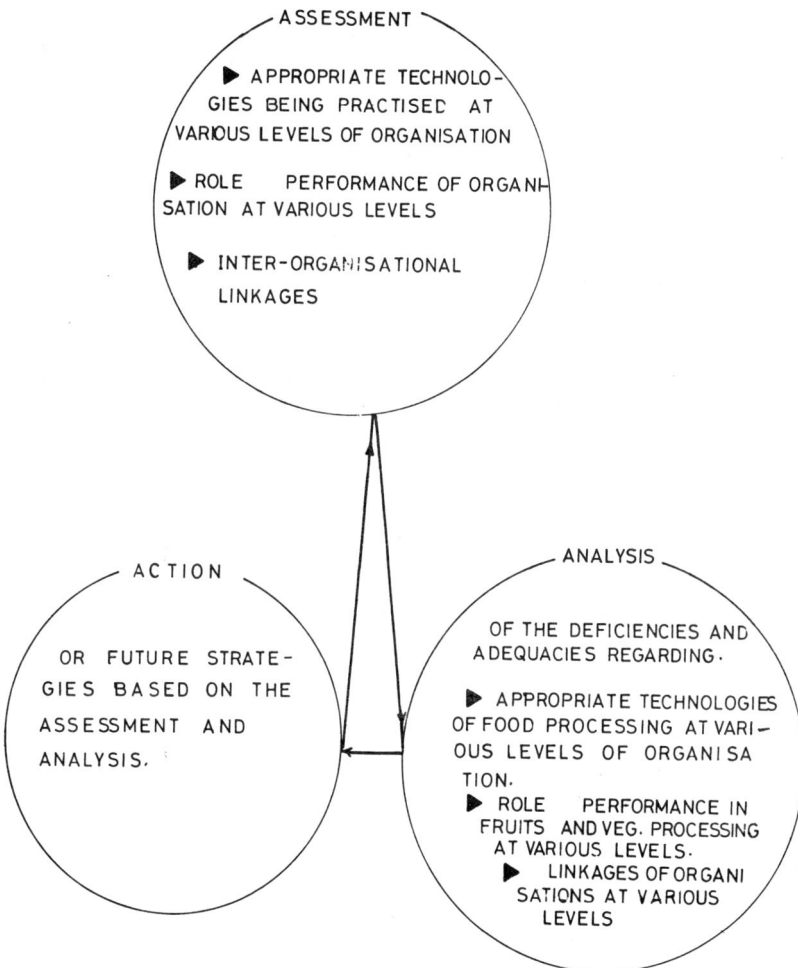

Fig. 5.14 : Triple 'A' Approach

Small Scale Units : Similarly, two small scale units were studied and both of them were using only one technology i.e., use of salt and oil.

Training Units : The training units were using three types of technologies like use of salt, use of sugar and use of chemical and other preservatives during various training courses organised for rural and urban women.

Khadi Gramudyog Unit : the Khadi gramudyog unit prepared the products with the use of sugar and salt.

Co-Operative Unit

Co-operative unit was also engaged in producting the products by the use of sugar, salt, chemical and other preservatives.

Use of salt is the main technology practised by various organisations.

ROLE PERFORMANCE

Large Scale Units

The production, training, research and development were the four major roles being performed by the Pan Foods ltd. Panipat while another large scale unit, namely, Pachranga International was performing production as its major role.

Small Scale Units

Both the small scale units had production as the major role.

Training Units : Training is the major role being performed by the training units. Besides training, they are also providing service facilities to the community and imparting nutrition education to the rural community.

Khadi Gramudyog Unit

Production and training are the two major roles of the khadi gramudyog unit.

Co-operative Unit

Co-operative unit is performing only the production role.

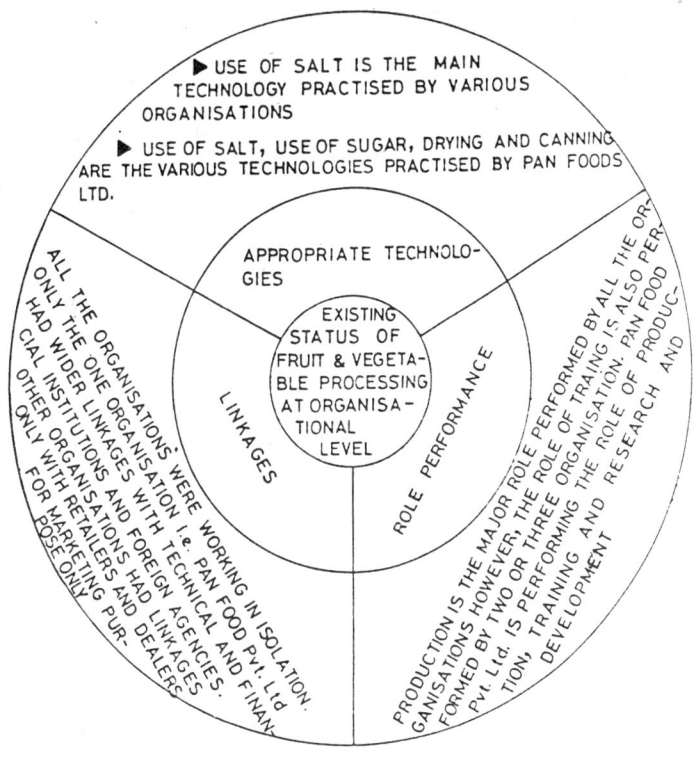

Fig. 5.15

Inter-Organisational Linkages

The study revealed that all the organisations were working in isolation. Only one organisation, namely, Pan foods Ltd. had a wider network of linkages with various institutes like technical/research institutes, financial institutes, foreign agencies and retailers and wholesalers for marketing purposes. Pachranga International and Murli Manohar fruit and vegetable products had linkages with foreign agencies that too for the export purposes. Training units are linked with technical institues, non-governmental organisations and various other departments like Women and child Development, Education, Health, Agriculture, Horticulture and Rural Development.

At institutional level, the proposed strategy presented here (Fig. 5.16) lays emphasis on promoting women's participation in fruit and vegetable processing activities at various levels of organisations. This developed strategy also constitutes three components, i.e., what, By whom and How.

Setting up large scale units at village level for providing employment to rural women and youth : Multipronged approach is needed to promote rural women's participation in fruit and vegetable processing. The one approach could be to encourage large scale investment by private and multinationals to enter this field, by way of tax concessi on and subsidies. In a developing economy like India, the growth of agro industries is significant for several reasons. First, it helps in generating more employment opportunities to rural people. A sound agro industry near and around the origin of agricultural surplus contributes to the market efficiency and helps in ensuring fair prices to the farmers. The major constraint is that rural people do not possess skills as required by the industry. But when the rural people get training from the technical institutes and training centres they should get employment opportunities near their village itself to restrict their influx to cities.

Thus, the agro industries have a good deal of potential in serving the rural people and improving their economy.

Developing improved technologies for fruit and vegetable processing : The constraints to rapid growth of food industry include : meagre data-base on production and processing, scarce scientific data pertaining to Indian situation, shortage and high cost of raw materials, infrastructural limitations, poor technology transfer-adoption system, and inadequate training facilities.

Strengthening of Research and Development (R&D) in foods is essential to enhance the technical capability of the nation, to generate, assess and absorb newer and more efficient technologies, to extend and improve food processing industry. Research and Development programmes should be planned by Agricultural Universities / Institutes, Horticultural research institutes / centres for taking up new research projects in the various areas which include dehydrated food, upgrading of traditional food technologies and design of new efficient and cost-effective processing machinery. Intensive efforts are also needed to strengthen the packaging capabilities for which R & D should take up desired investigations.

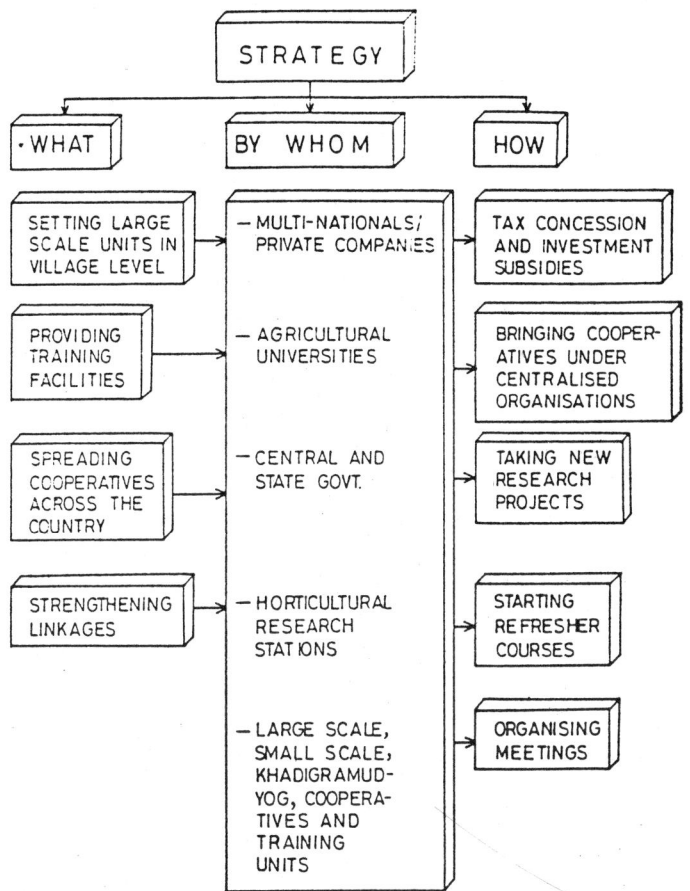

Fig. 5.16 : Strategy for promoting women's participation in fruit and vegetable processing at organisational level.

Providing Training Facilities : Human resources development is an integral part of industrial development. There are few courses being run by some universities. The quantity and quality of personnel coming from these institutions are not considered adequate. To overcome the problem of quantity and quality of personnel at various levels of organisations agricultural universities should start a programme to train food engineers to fill the vaccum in the food processing industry. Technical institute should start 15-day refresher courses for industry personnel and impart upto-date research information to them.

Strengthening Linkages : There is a very simple explanation why despite nature's abundant gifts Indian fruit and vegetable processing industries perform so poorly in making use of them to bring a little warmth in the life of the millions of people in the rural areas who are involved in the processes of cultivating, transporting, processing and consuming. In order to contain the increase in production costs, various steps should be taken. One of the steps is to link growing and processing in the form of contract farms attached to processing units or through direct contacts with a large number of growers by processors and preservers. Secondly, a large number of extension staff exists. They should be trained and atleast asked to assist the processed food industry because they have much closer linkages with the farmer.

Another kind of linkage can be in the form of computerisation support from the large industry to the small industry. If large scale sector have their own operation, and they are also having a computerised programme for their operations in small measure. They could develop a package and common cost could be spread over the franchisees and small units.

A complete kind of linkage between the small and large sector is one in which the large sector provides every possible input to the small scale units to upgrade the variety of the product, the quality, technology and the quality of the people in terms of training inputs. The large sector can advise small sector units on the four major essentials. They are product standardisation, product presentation, product placement and product prmotion. ¯

The principal missing link in Indian fruit and vegetable processing industry is the lack of an integrated technological approach but not the processing technology.

Encouragement should be given to farmers to form cooperative societies among fruit and vegetable growers in horticulturally rich areas. These societies contract with processing units, whether these are in private or in the cooperative sector or in the public sector. These types of cooperatives should spread across the country and may be revitalised to play a much larger role. This could be achieved only if they are all brought under a centralised organisation so that they move from 'regional' to the 'national' level operation.

6

SUMMARY AND CONCLUSION

The Green Revolution and subsequent efforts through the application of science and technology for increasing food production in India have brought self-reliance in food. The impetus given by the Government, State Agricultural Universities, State Departments of Agriculture and other organisations through the evolution and introduction of numerous hybrid varieties of cereals, legumes, fruits and vegetables and improved agricultural practices have resulted in increased food production. In the post-green revolution era, eventhough food grains have been taken care of, fruits and vegetables for want of simple technologies of processing, preservation and transport to various places of need have suffered post harvest losses, estimated to be nearly 35 per cent. Today only one per cent of the total fruits and vegetables are processed in the 3,000 food industries in the country.

Therefore, food processing has been engaging the attention of planners and policy makers as it can contribute to the economic development of rural population. Since women constitute 50 per cent of population, they play an important role in the economic upliftment of the family. Traditionally, women handle food and are familiar with the skills of food processing. In order to improve the status of women and rural food processing, low cost indigenous food processing technologies offer excellent opportunities for women in production of processed foods.

It is difficult to exactly quantify the role of rural women in fruit and vegetable processing. But there is a need of having a fuller understanding of the role and contribution of women in fruit and vegetable processing so that extension services may accordingly be attuned to fully integrate them in food processing. The study of different linkages seems to be necessary to gather this requisite information. For this reason, the present study was proposed with the following objectives :

1. To study the role performance and linkages of different organisations and institutions working for fruits and vegetable processing.

2. To identify the appropriate food processing technology for dissemination among the rural families.

3. To study the crucial factors influencing food processing at different levels.

4. To suggest the strategy for promotion of participation of rural women in food processing.

The study was conducted in Hisar, Panipat and Faridabad districts of Haryana. Two large scale and two small scale organisations were selected randomly whereas two training organisation, one khadi gramudyog and one cooperative were selected purposively for case study.

Sixty five rural families, working in selected organisations and engaged in fruit and vegetable processing, either at commercial level or home scale, were selected purposively.

To study the role performance of organisation's personnel, the personal, psychological and organisational variables were studied. Role performance and linkages were dependent variables for this study. The antecedent variables for rural families were socio-personal, psychological and resource system. The identification of the appropriate technology for fruit and vegetable processing was done by judge's consensus, secondary data and by conducting experiment.

The investigation was conducted through different types of well structured interview schedule. The data were analysed through standard statistical techniques such as percentages, mean, 't' test, correlation coefficients and multiple regression.

The important findings of the present investigation are briefly summarised as under :

Role performance and linkages at various levels of organisations and institutions : The study revealed that production was the major role being performed by various organisations. Only one organisation, namely, Pan food ltd. dealt with the roles of production, training, research and development. Training units played major role in training rural and urban women. Besides training, they also offer service facility to needy community and impart nutrition education to create awareness among rural people.

The organisations at different levels were engaged only in the production of pickles. Only one organisation, namely, Pan Food Pvt. Ltd. produced various products like pickles, jams, jellies, soups, syrups, squashes, murabba, fruit and vegetable powder, canned mushroom, chutney, sauces, ketchup, etc. As regards the use of processing technology, most of the organisations were found to be using only one technology, i.e., use of salt. Some of them used sugar for preparing the products. However, Pan Food Ltd. employed several technologies, i.e., salt, sugar, chemical, other preservative, drying and canning for manufacturing of various products.

Comparatively less women than men were employed at various levels of organisations. Women were engaged only for manual work like cleaning of spices, bottles, cans, fruits, vegetables, etc., while men were engaged for mechanical jobs.

Attitude towards the job by majority of personnel of large scale, small scale, khadi gramudyog and cooperative units was moderately favourable.

In large scale units about 48.7 per cent of the respondents were satisfied with their jobs, whereas in small scale and training units 50 per cent of the personnel had moderate level of satisfaction. Almost one hundred per cent personnel of khadi gramudyog enjoyed moderate satisfaction.

Satisfaction about organisational climate perceived by majority of personnel at different levels was found to be moderate. And, in training and cooperative units, the same was satisfactory.

Organisational commitment in the personnel of cooperative unit was high and was low in the majority of personnel of large scale, small scale, training and khadi gramudyog units.

Professional commitment was high in the majority of personnel of large scale, training and khadi gramudyog units whereas it was moderate in most of the personnel of small units.

Major constraints faced by various organisations were limitations of lower staff, lack of funds, competition from established units and difficulty in getting money from buyer after sale whereas administrative constraints were faced only by the training units.

All the organisations were working in isolation. Except for two organisations none of the them had any linkage with technical institutes, fruit and vegetable growers of the neaby villages and governmental and non-governmental organisation. Pan Foods Ltd. had wider linkages with technical institutes, financial institutes and foreign agencies.

Most of the respondents of rural families were of old age, belonging to nuclear family and had medium sized family. Majority of them were of middle caste, family education status and illiterate. Most of them were not the members of any social organisation. Majority of the men were having frequent urban contact. It was found that majority had no extension contact. Most of the women were somewhat economically motivated, less risk-oriented, having moderately favourable attitude and low level of knowledge. Most of the rural families were doing the fruit and vegetable processing at commercial level. All (100%) of the families were practising the technology of drying and use of salt with oil.

Maximum performance of women was in identifying need for inputs, establishing processing priorities and modifying plans and low performance in making contacts for finances, purchase of raw material and arrangement of finances in managerial roles.

At institutional level also, maximum performance of women in processing role was in activities like washing, selecting and blanching and low in packaging, preserving and chopping.

With regard to marketing roles, maximum performance of women was in seasonal planning and checking the prices.

In managerial roles men performed maximum in activities like arrangement of finances, making contacts for finances and purchase of raw material.

As regards the processing roles, maximum performance of men was in packaging, preserving and chopping.

In marketing roles men were performing mainly for promotion of products, checking the price of products and planning marketing according to season.

It was found that 55.38 per cent of women performed poorly in managerial roles in comparison to men (56.93%) performing good.

For processing roles, only 15.39 per cent of men and women performed good, whereas 84.61 per cent women had an average performance.

As regards marketing roles, majority of the women had poor performance in comparison to 72.31 per cent of men performing good.

Managerial, processing and marketing roles performed by men and women in fruit and vegetable processing showed significant difference between them.

The major constraints faced by rural families were lack of appropriate technology, dual responsibilities, lack of funds, getting money from buyer after sale and availability of raw material at a high price.

Very less number of respondents had linkages with families of their own village, other villages and technical institutes. None of the rural families had linkages with financial institutes.

Identification of appropriate technology of fruit and vegetable processing for dissemination in rural areas : On the basis of judges consensus, the evaporative cooling system/zero energy cool chamber for post-harvest stage, drying for processing stage and tray packing emerged as the appropriate fruit and vegetable processing technology for dissemination in rural areas.

Zero energy cool chamber and drying emerged appropriate suitable technology of fruit and vegetable processing from the review data.

It was found by conducting comparative study on natural convention solar dryer (Chimney) and open sun-drying, indirect sun drying took more time to dry vegetables but the quality of the product was good in comparison to open sun drying.

Factors influencing food processing at institutional level: It was revealed that education, caste, urban contact, type of unit, availability of raw material, training and machinery and equipment possession had significant positive correlation, whereas constraints had negative relationship with role performance.

Economic motivation, type of unit, availability of raw material, machinery and equipment possession had significant, positive correlation with the linkages of rural families.

Education, caste, family urban contact, type of unit, availability of raw material, training, machinery and equipment possession and constraints together explained 90 per cent of variation in the performance of roles in fruit and vegetable processing. Individually, type of unit exhibited significant regression coefficients towards role performance.

Economic motivation, type of unit, availability of raw material, machinery and equipment possession had significant positive correlation withthe linkages of rural families.

Education, caste, family urban contact, type of unit, availability of raw material, training, machinery and equipment possession and constraints together explained 90 per cent of variation in the performance of roles in fruit and vegetable processing. Individually, type of unit exhibited significant regression coefficients towards role performance.

Type of unit, availability of raw material, economic motivation and machinery and equipment possession together explained 61 per cent variation in the linkages of rural families. Individually, type of unit exhibited significant regression coefficients towards linkages of rural families.

Conclusion : In nutshell, the study revealed that at organisational level production is the major role and almost all the organisations except one were using only one technology for the production of products. At different organisational levels women were employed less in number as compared to men and were engaged only for mannual jobs. Inter-organisational linkages were negligible. Only one organisation, namely, Pan Food ltd. had a wide network of linkages, whereas other organisations had linkages with dealers and retailers for marketing purposes.

At the institutional level the women's performance in managerial and marketing roles was poor in comparison to men. There was a significant difference between the performance of managerial, processing and marketing roles of men and women.

Linkages at institutional level were negligible with technical and financial institutes. Only few families had linkages with families of their own and other villages who were also engaged in fruit and vegetable processing activities.

For dissemination of technology in the rural areas, drying emerged as an appropriate fruit and vegetable processing technology.

Suggestions : Suggestions on removal of poor participation, inadequacies of women in food processing industry and poor performance at institutional level have been incorporated in the strategy for promoting women's participation in fruit and vegetable processing organisations and institutions. However, a few seggestions for future research are :

1. To study in detail the roles and activities of organisations, their personnel and rural families through investigator's Personal Observation (IPO) Technique.

2. All input and output factors need to be studied in detail to determine economics of production with a special reference to making a Food Processing Enterprise remunerative for rural development.

3. Cost-effective and adoptable technologies need to be developed and tried for their acceptability, productivity and production efficiency.

4. To develop modules of monitoring system to determine intensity and extensity of consumption of different food products for structuring and developing a viable marketing system.

BIBLIOGRAPHY

Allport, G.W. 1935, Personality : A Psychological Interpretation. Henry Halt & Co. New York.

Anonymous 1983. Annual report, CFTRI

Anonymous. 1986. Indian Fd. Packer. 40 (5) : 58.

Anonymous. 1986. Indian Fd. Packer 40 (5) : 59.

Anonymouus. 1986. Solar Crop drying- TDRI. Indian Fd. Packer. 40 (6) : 72-73.

Anonymous. 1988. Problems of food processing industry. Financial Express. December 3.

Anonymous 1988. Financial Express. December 29.

Anonymous. 1990. Industry agriculture linkages. Indian Farmer Times. 8 (9) : 24-25.

Anonymous. 1990. Industry agriculture linkages will help development. Indian Farmer Times. 8 (9) : 11-13.

Anonymous. 1991. Indian Fd. Packer. 45 (3) : 7-8.

Anonymous. 1991. Indian Fd. Packer. 45 (8) : 53.

Anonymous. 1992. Food processing industries marching ahead. Indian Express. January 15.

Anonymous. 1993. Industry-agro units linkage mooted. Economic times. Feb. 18.

Bagchi, D.P. 1986. Indian Fd. Packer. 40 (3) : 27-29.

Bennet, E.J. and Tumin, K. 1948. Human Society, Mac millian, New York.

Bhople, R.S. and Patki, Alka 1992. Correlates of role

performance and training needs of farm women labour. Journal of rural development. **11** (1) : 49-58.

Bose, S. 1995. Caste, tribe and female labour participation. Social change. **14** (2) : 15-20.

Chakravarty, S. 1975. Women power in agricultural development. Kurukshetra. **24** (4) : 7.

Chandel, S.S. 1992. Solar dryer propagation in cold areas of Himachal Pradesh. Paper presented at National solar energy convention held at new Delhi, 28-30th December.

Chaturvedi, Anurag. 1990. Low cost technologies for preservation of vegetables. Paper presented in summer institute on appropriate food processing technologies for rural development 15th June - 4th July, Hyderabad.

Chaudhary, J.L. 1989. Agro based and food industries-A Review. Khadi gramodyog **36** (10) 61-69.

Das, A.K. 1991. Policy Issues on Processed Food Industry. Indian Food Industry, **10** (1): 14-25.

Davis, Kingsley. 1949. Human society. The Mcmillan co., New York.

Devadas, B. and Sikidar, Surjit 1990. Development of small and medium enterprises : Need for entrepreneurial structure in a developing economy. Proceedings of eight national convention of women entrepreneurs. Organised by national Alliance of Young Entrepreneurs, New Delhi, pp. 302-306.

Devadas, Rajamal, P. 1980. Rural processing for better utilisationof fruits and vegetables. Social welfare. **27** (6) : 37-39.

Devi, L. 1983. Role expectations and role performance of rural women in farm and home management. Ph. D. thesis (Unpub.), A.P.A.U. Hyderabad.

DGTD Report 1984. In Indian Farmer Times. **7** (1): 21-22.

DGTD. 1989. The newly emerging industry of food processing. Indian Farmer Times. **7** (2) : 27-32.

Dhabade, R.S. and Khedkar, D.M. 1980. Part IV. Indian Fd. Packer. **34** (3) : 35.

Economic Times. 1992. Indian Fd. Packer **46** (5) : 10.

Edwards, A.L. 1957. Techniques of attitude scale construction. Appletoncentury -Crafts : New York.

Garrett, H.E. 1979. Statistics in psychology and education. Vakils, Feffer and simons Ltd., Bombay.

Geervani, P. 1990. Appropriate food processing technologies and role of home science faculty. Paper presented in the summer institute on appropriate food processing technologies for rural development. 15th June-4th July, Hyderabad.

Ghosal, M.K. 1990. Food processing : Thrust area for export and foreign collaboration. Yojana **34** (10) : 11-12.

Guilford, J.P. and Benjamin Fruchter. 1956. Fundamental statistics in psychology and education. MC Graw-Hill Kogakusha Ltd., Tokyo.

Habibunnisa, Edward Aror and Narasimham, P. 1988. Extension of the storage life of fungicidal waxol dip treated apples and organges under evaporative cooling storage conditions. J.Fd. Sc. Technol. **25** (2) : 75-77.

Harigopal, U and Tonapi, K.V. 1980. Technology for villages : Solar drier. Khadi gramodyog. **26** (11) : 491-493.

Harigopal, U. and Tonapi, K.V. 1981. sun drying of palm kernel. Indian Fd. Packer. **35** (1) : 27-28.

Hindustan Times. 1990. Indian Fd. Packer. **44** (4): 52.

Hindustan Times. 1990. Indian Fd. Packer. **44** (6) : 65.

Hindustan Times 1992. 10th November.

Jain, S.P. 1992. Inter-organisational linkages at village level : a study. Journal of rural development **1** (2) : 271-306.

Jauch, L.R., Glueck, W.E., osborn, R.N. 1978. Organisational loyality, professional commitment and Academic Research Productivity. Academy of Management Journal, **21** (1) : 84-92.

Jayaraman, J.S., Das Gupta. D. K. and Babu Rao, N. 1991. Quality Characteristics of some vegetables dried by direct and indirect sun drying . Indian Fd. Packer. **45** (1) : 22-26.

Joshi, A.R. Agarwal, Devraj; Singh, Gunveer; Kakkar Monika; Prasad, P.M. and Ray, U. 1989. Backward linkages of the food processing industry in India. Indian Fd. Packer. **43** (2) : 25-33.

Kadam, K.R, and Valuj, D.R. 1982. Role performance of gram panchayat members in village development activities. Mah. J. of Ext. Edu. **1** (1) : 37-42.

Kalra, S.K. and Bhardwaj. 1981. Use of simple solar dehydrationfor drying fruits and vegetables products. J. Fd. Sci. Tecnol. **18** : 23-26.

Kaur, S. 1991. Role of farm women in selected agricultural operations in five villages of Ludhiana district. M.Sc. thesis (Unpub.), Punjab agricultural university, Ludhiana.

Kaur, s. 1986. Women in rural development. A case study. Mittal publications, Delhi.

Khare, H.C., Jaiswal, D.K. and Mishra, P.K. 1987. Role performance of S.M.S.S. in T and V extension System. Mah. J. of Ext. Edu. **6** : 199-200.

Kherde, R.L. and Sahay, B.N. 1972. Role performance and role prediction of the village level workers in the new strategy of agricultural production. Indian J. of Ext. Edu. **8** (2) : 67-70.

Khurdiya, D.S. and Roy, S.K. 1986. Solar drying of fruits and vegetables. Indian Fd. Packer. **40** (4) 28-39.

Krishnaswamy, P.K. 1987. Research and development-Scope and limitations. Indian Fd. Packer. **41** (3) : 71-72.

Kumari, Sarala, A. 1990. Low cost technologies for preservation of fruits. paper presented in Summer institute on appropriate food processing technologies for rural development 15th June-4th July, Hyderabad.

Kunwar, Narayan and Willams, D.L. 1990. Performance of tasks by field level agricultural extension workers in Nepal. Indian J. of Ext. Edu. **26** (1&2) : 1-12.

Kurade, A.G. Naik. 1991. Marketing of processed fruits and vegetables in India. Indian Fd. Packer. **35** (2): 112-119.

Lacke, E.A. 1976. the nature and causes of Job satisfaction in Dunnette, M.D. (ed). Handbook of Industrial and Organisational Psychology. Rano Mcnally; Chickago.

Laharia, S.N. 1978. A study of personal and organisational variables influencing the productivity of agricultural scientists. Ph. D. thesis, Haryana Agri. Univ., Hisar.

Likert, R. 1932. A technique for the measurement of attitudes. Arch. Psychol., No.140.

Mahadeviah, M., Gowramma, R.V. and Naresh, R. 1990. Studies on the suitability of tinplate with reduced tincoating for packing fruit and vegetable products. Indian Fd. Packer. **44.** (6) : 25-36.

Maini, S.B., Anand, J.C., Kumar, Rajesh, Chandan, S.S. and Vishishith, S.C. 1984. Evaporative cooling system for storage of potato. Indian J. Agric. Sc. **54** (3) : 193-195.

Maini, S.B., Diwan, Brijesh, Gupta, S.K. and Anand, J.C. 1984. An inexpensive field solar dryer for fruits and vegetables. Indian Fd. Packer. **38** (6) : 77-79.

Maini, S.B., Diwan, Brijesh, Lal, B.B. and Anand, J.C. 1984. Comparative performance of packing apples in trays and conventional pack during transit. J. Fd. Sci. and Tech. **21** (6) : 409-410.

Maini, S.B., Diwan, Brijesh and Anand, J.C. 1985. Studies on the comparative efficiency of solar dryers for drying potato chips. Indian Fd. Packer. **39** (2) : 25-28.

Malaviya, M.K. Gupta, R.S.R. and Tyal, G. 1987. solar dryer for drying fruits and vegatables for use in rural area. Proceedings of Role of food technology in rural development, 142-147.

Malviya, M.K. and Yadav, Y. K. 1992. Comparative performance testing of low cost solar driers. Paper presented at national energy convention, held at new Delhi, 28-30 Dec.

Mandhyan, B.L., Abrol, C.M. and Tyagi, H.R. 1988. Dehydration characteristics of winter vegetables. J. Fd. Sc. Technol. **25** (1) : 20-22.

Maninck, T.N. and Manimegali, G. 1988. Appropriate Home Science Technologies for the upliftment of rural women. Paper presented in International Conference on appropriate agricultural technologies for farm women, held at I.C.A.R., New Delhi.

Mishra, P.K. Jaiswal, D.K. and Mishra, Anupam. 1988. Factors affecting role performance of RAEO working in Jabalpur district. Mah. J. of Ext. Edu. **7** : 231-243.

Moulick, T.K. 1965. A study of the prediction values of some factors of adoption of nitrogenous fertilizers and the influence of sources of information on adoption behaviour. Ph.D. Thesis. I.A.R.I. New Delhi.

Mukherjee, S. 1990. Rural food handling and processing industries. paper presented in summer institute on appropriate food processing technolgies for rural development. 15th June-4th July, Hyderabad.

Mukhopadhyay, T.K. 1981. Food processing and its relevance to rural development. Indian Fd. Packer **35** (1) : 31-38.

Naik, P.H. 1989. Development of entrepreneurs in villages. Khadi gramudyog. **36** (3) : 159.

Narwal, R.S. 1982. Changes in farmers knowledge about and attitude towards some selected aspects of irrigation water in a traditionally dry farming tract of Haryana. Ph. D. Thesis, Haryana Agri. Univ., Hisar.

Nikhade, D.N. and Kitey, P.V. 1984. Role performance of village level workers in agricultural development programme. Mah. J. of Ext. Edu. **3** 111-112.

Noll, V.H. 1957. Introduction to educational measurement. Boston, Hanghton Miffin Company.

Pandey, U.B. 1989. Raw material status and processing industry. Indian Fd. Packer. **43** (4) 73-80.

Pandit, A.R. 1983. Backward linkages of food processing industry. Indian Fd. Packer. **37** (3) : 11-15.

Patil, Shivraj 1984. Government stands for removal of obstructions. Indian Fd. Packer. **38** (3) : 9-10.

Pawar, V.N., Singh, N.L., Devi, D.K., Kulkarni, D.N. and

Ingle, U.M. 1988. Solar drying of white onion flakes. Indian Fd. Packer. **42** (1) : 15-28.

Pestonjee, D.M., 1973. Organizational structures and job attitudes. The Minerva Associates : Calcutta.

Porter, L.W., Stress, R.H., Mow Day, R.T. and Boulian, P.V. 1974. Organisational commitment questionnaire c,f, Halold L. Angle and James, L.R. (1981). An empirical assessment of organisational commitment and organisational effectiveness. Administrative Science Quarterly, **26** (1) : 1-13.

Prahlad, S.N. 1992. Emerging trands in the fruit & vegetable processing industries industry on University cooperation research. Indian Fd. Packer, 46 (6) : 32-34.

Prasad, Kamala, 1981. Indian Fd. packer. **35** (2) : 88-90.

Rade, V.M., Desai, B.R., and Girase, K.A. 1991. Role perception and role performance of contact farmers in T & V system. Mah. J. of Ext. Edu. **10** (2) : 154-157.

Raghunandan, A. 1992. Diversifying agriculture by boosting horticulture. financing Agriculture. **24** (4) : 9-12.

Rao, S.V.N. and Sohal, T.S. 1985. Improving the performance of veterinary surgeons. Mah. J. of Ext. Edu. **4** : 19-26.

Ray, D.M. 1990b. Strategic implications to entrepreneurial venture born international four case study. Proceedings of ENDEC International enterpreneurial conference, held at singapore, March 21-26, PP. 338-343.

Reddy, S.K. and Mulay, S. 1972. Differential performance of leadership roles in two north Indian village communities. Indian J. of Ext. Edu. **8** (1): 7-13.

Rege, D.V. 1992. What ails the Indian Food industry. Indian Fd. Packer. **45** (5) : 7-12.

Rizvi, R.S. 1967. Job analysis, job performance and suitability of pre-service training of gram sevikas in three selected states. M.Sc. Thesis, I.A.R.I., New Delhi.

Roy. S.K., and Khurdiya, D.S., 1986. Studies on evaporatively called zero energy input cool chambers for the storage of horticultural product. Indian Fd. Packer. **40** (6) : 26-31.

Roy, S.K. 1991. Development of infrastructural facilities to boost the export of fresh and processed horticultural produce. Indian Fd. Packer. **45** (5) : 13-18.

Sarin, K.P. Theme paper. Indian Fd. Packer **46** (10) : 59-63.

Sethi, V. and Maini, S.B. 1989. Appropriate technology for reducing post harvest looses in fruits and vegetables. Indian Fd. Packer. **43** (2) : 42-56.

Sethi, Vijay, Jai Bhagwan, Behal, Neeta and Lal, Shashi. 1991. Low cost technology for preserving mushrooms. Indian Fd. Packer. **45** (6) : 22-26.

Sharma, M.L., Sharma, P.N., Jaiswal, P.K. and Sengar, R.S. 1988. Role expectation and role performance of rural agricultural extension officers of T & V system in Madhya Pradesh. Indian J. of Ext. Edu. **24** (3 & 4) : 75-78.

Shearer. A.K. 1961. General psychology. Berkeley university of California press, California.

Singh, Kamaljeet, 1984. Bright future for processed fruits and vegetables. Indian Fd. Packer. 38 (3) : 11-15.

Smith, H.C., 1955. Psychology of Industrial behaviour, Mcgraw Hill : New York.

Snedecor, G.W., and cochran, W.G. 1968. Statistical methods. Ames Iowa state University Press.

Sohi, J.S. and Kherde, R.L. 1986. Job performance and job prediction of the livestock assistants as perceived by their supervisors. Mah. J. of Ext. **4** : 59-66.

Srivastava, K.B. 1982. Management in rural development : A study of linkages among panchayat samitis, cooperatives and voluntary organisations. Journal of rural development. **1** (5) : 706-736.

Supe, S.V. 1969. Factors related to different degree of rationality of decision making among farmers. (Unpublished), Ph.D. Thesis, I.A.R.I., New Delhi.

Tiwari, N.D. 1981. Indian Fd. Packer. **35** (2) : 102-104.

Trivedi, G. 1963. Measurement analysis of socio-economic status of rural families. (Unpublished), PH.D. Thesis, I.A.R.I., New Delhi.

Trivedi, S. 1979. Women vital role in rural development, Rekha printers Pvt. Ltd., New Delhi.

Vaghani, S.N. and Chundawat, B.S. 1986. Solar drying of Sapota fruits. Indian Fd. Packer. **40** (2) : 23-28.

Varde, S.D. 1991. Indian Fd. Packer (Theme Paper) **45** (5) : 59-60.

Varma, U. 1987. An analysis of communication pattern among information generating, information disseminating and information utilizing systems of Home Science in Haryana. Ph.D. Thesis, Haryana Agri. Univ., Hisar.

Vyas, D.M. and patel, J.C. 1991. Rural development through food processing. Yojana. **35** (7) : 6-7.

Waris, Amtul, Reddy, M.N. and Anjanappa. M. 1990. Role performance and job satisfaction of the anganwadi workers of I.C.D.S. in Andhra pradesh. Indian J. of Ext. Edu. **26** (1 & 2) : 119-121.

Xavier, M.J. and Gopalaswamy, T.P. 1992. Fruits and vegetable marketing : Concerted efforts vital. Financial express. Nov. 23.

Yadav, R.Z. and Azad, M.P. 1987. Role of women in allied enterprises for rural development. Kurukshetra. **36** (2): 15-18.

Yadav, S.S. Srivastav, V.K. and Kumar, Virender. 1989. Post-harvest management in vegetable crops. Indian Farmer Times. **6** (11) : 15-16.

ANNEXURE-I

List of fruit and vegetable processing organisations of the selected Distiricts.

DISTRICT PANIPAT

Category	Name of Organisation
Pvt./SS	M/s Bharat Achar Factory (9, Bishan Sarup Colony), Panipat
Pvt./CS	M/s Hari Shankar & Sons. (DV/697/40, Sector 8, Khel Bazar
Pvt./SS	M/s Murliwala Fruit Products (E-73, Industrial Area)
Pvt./LS	M/s Pan Foods Ltd. (G.T. Road)
Pvt./RL	M/s Pan Foods Ltd. (G.T. Road)
Pvt./RL	M/s Murlidhar Ram Kishan (225-R, Model Town)
Pub./HS	State Institute of Catering and Hotel management
Pvt./LS	M/s Pachranga International (Behind Rajwoolen Inst.)
Pvt./RL	M/s Pachranga International
Pvt./SS	M/s Murlidhar Ram Saran (13, Nehru Nagar)
Pvt./RL	M/s Murlidhar Ram Saran (Pachranga Bazar)
Pvt./HS	M/s Hemant Fruit Product (Bhatia Colony)
Pvt./RL	M/s Pacranga Acharwala (3257, Behind Inderpuri)

Category	Name of Organisation
Pvt./LS	M/s Pachranga Foods (C-3, Industrial Area)
Pvt./RL	M/s Pachranga Foods (-do-)
Pvt./SS	M/s Pachranga Goods (604, Pachranga Bazar)
Pvt./HS	M/s Ramesh Jindal Achar Factory (47/5, Moh.Bari Pahar)
Pvt./RL	M/s Murlimanohar & Food Products (236, Model Town)
Pvt./RL	M/s Pachranga Manufacturer (417R, Model Town)
Pvt./RL	M/s Pachranga Manufacturer (417R, Model Town)
Pvt./SS	M/s Murli Manohar Fruit Products (331, Model town)
Pvt./RL	M/s Murlimanohar Foods
Pvt./RL	M/s Pachranga Overseas (605/4, Pachranga bazar)

DISTRICT HISAR

Coop/CS	M/s Hisar Ideal Fruit and Vegetable Product Coop. (Near brahma Vidyalya)
Pvt./SS	M/s Pooja Achar Factory (83, Near New Sabzi Mandi)
Pvt./SS	M/s Chabra Achar Factory (409/10, Gossin Wala Mohalla)
Pvt./HS	M/s New Bharat Products (Bhambaram Colony, Fatehabad)
Pvt./HS	M/s RAju Achar Murabba Gramudyog Mandal (83, Sant Nagar)
Pvt./HS	M/s Arpan Food Products, Hansi (4713 RAmpura)
Pvt./HS	M/s Kapil Fruit Product (207, New Model Mandi)
Pvt./CS	M/s Rinku Drinks (Village SAtrod)
Coop/SS	M/s Nirmal Drink (50/11, Dhani Barwali)
Pvt./LS	M/s Skippeo Beverages Pvt.Ltd. (Mile Stone 3-6 KM Hansi-Hisar Road)
Pvt./SS	M/s Haryana Phal Nigam (Satrod Inds. Area)

Category	Name of Organisation

DISTRICT FARIDABAD

Pvt./HS	M/s Haryana Cold Drinks (3C/79 B.P. NIT)
Pvt./LS	M/s Agro Fab. Industries (P) Ltd. (15/1, Mathura Road)
Pvt./HS	M/s Durg Udyog (14/15 B, Bangalow Plot, NIT)
Pvt./HS	M/s NIT's Food and Bev. (B-190, NIT, Nehru Ground)
Pvt./HS	M/s A.A. Associates, Faridabad.

ANNEXURE-II

HARYANA AGRICULTURAL UNIVERSITY
HISAR-125004

Dr./(Mrs,) U.Varma,
Assoc. Prof. & Head,
Depptt. of Home Science
Extension Education   Dated.....................

Dear Sir/Madan

Ms. Suman Bhatti, a Ph.D. student of this Department has undertaken a research project entitled "Role analysis and linkages of fruit and vegetable processing organisations and institutions by rural families", of the several variables included in the study, measurement of attitude of organisation's personnel for their job and rural families attitude towards fruit and vegetable processing which entails the preparation of a scale. Therefore, the statements given have to be rated for their relevance for inclusion in a particular scale. On a 5 point continuum ranging from 'Strongly agree to strongly disagree'.

Effort has been made to prepare an exhaustive list of pertinent statements, but you have scope to add as many more statements as you may deem fit. Likewise, you also have the option to modify or restructure any of the statements.

You are requested to kindly determine the relevancy of each item with respect to kindly determine the relevancy of the scope of definition given by Beal and Sibley, 1967 which is as under:

Attitude : Based on the values system man develops attitude the relatively enduring sets of positive or negative evaluation, emotional feeling and pros and cons tendencies to act towards physical or social object.

Please do keep in mind that you are acting as a judge in your capacity as a professional expert and your own attitude has not to be projected. I am confident that you will devote your valuable time to this work.

With regards,

Yours sincerely,

Sd/-

(U. Varma)

Reference

Beal, G.M. and Sibley, D.N. (1967). Adoption of Agricultural Technology by the Indians of Guatemala. Rural sociology Report-62, Ames, Iowa State University of Science and Technology.

ANNEXURE-II (A)

ATTITUDE STATEMENT FOR ORGANISATIONS PERSONNEL IN RELATION TO THEIR JOB.

Sr. No.	Statements	Degree of agreement/ disagreement				
		SA	A	UD	DA	SDA
1.	I consider my job as a noble job					
2.	To me nothing is more satisfying than working in fruit and vegetable processing.					
3.	I take my job as one of my moral duties.					
4.	I always try to accept changes for better perfor-mance.					
5.	The sooner I get out of the job in fruit and vegetable processing in-dustry, the better it is.					
6.	Working in fruit and vegetable processing is very important for natio-nal development.					
7.	Working in fruit and vegetable processing is the need of time					

Sr. No.	Statements	Degree of agreement disagreement				
		SA	**A**	**UD**	**DA**	**SDA**
8.	Working in fruit and vegetable processing is not rewarding.					
9.	Working in fruit and vegetable processing is very satisfying for me.					
10.	There is no use of working in fruit and vegetable processing organisation as there is no chance of making money.					
11.	I have indentification with my job.					
12.	I consider my job challenging in the era of technological advancement					
13.	I try to put my head and heart into my job without bothering about recognition or promotion					
14.	I will like to opt for another job even at lesser salary.					
15.	I have control over my feelings to avoid job stress.					

ANNEXURE-II (B)

ATTITUDE STATEMENTS FOR RURAL FAMILIES TOWARDS FRUIT AND VEGETABLES PROCESSING

Sr. No.	Statements	Degree of agreement/ disagreement				
		SA	A	UD	DA	SDA
1.	One should start fruit and vegetable processing as there is vast scope of economic benefit.					
2.	Education and training can easily solve the problems of less knowledge regarding fruit and vegetable processing.					
3.	Starting fruit and vegetable processing is time consuming, therefore one should not waste time on it.					
4.	Starting food processing is money consuming, therefore one should not waste money on it.					
5.	How can one well ensure economic benefit by starting it.					

Sr. No.	Statements	Degree of agreement/ disagreement				
		SA	A	UD	DA	SDA
6.	Starting fruit and vegetable at commercial level is feasible in city area therefore, rural people should not start it.					
7.	The knowledge of food processing is beyond the comprehension of illiterate and unskilled people, therefore one should not start it.					
8.	As the cost of raw material is high so it is not feasible for rural people.					
9.	No one should agree with economic benefits by starting it.					
10.	There is no harm in starting food processing because it promotes better living.					
11.	If one has enough knowledge of food processing it's easy to start even without raining.					
12.	I try to accept changes and methods in food processing for better performance.					

ANNEXURE-III (A)

SELECTED ATTITUDE STATEMENTS FOR ORGANISATION'S PERSONNEL IN RELATION TO THEIR JOB.

Sr. No.	Statements	't' values
1.	I consider my job as a noble job	0.98
2.	To me nothing is more satisfying than working in fruit and vegetable processing organisation.	1.92*
3.	I take my job as one of my moral duties.	1.63
4.	I always try to accept changes for better performance.	1.82*
5.	The sooner I get out of the job in friut and vegetable processing industry, the better it is.	1.95*
6.	Working for fruit and vegetable processing is very important for national development.	1.76*
7.	Working for fruit and vegetable processing is the need of time.	2.10*
8.	Working in fruit and vegetable processing is not rewarding.	1.92*
9.	Working in fruit and vegetable processing is very satisfying for me.	1.72*

Sr. No.	Statements	't' values
10.	There is no use of working in fruit and vegetable processing organisation as there is no chance of making money.	1.94*
11.	I have identification with my job.	1.12
12.	I consider my job challenging in the era of technological advancement.	2.22*
13.	I try to put my head and heart into my job without bothering about recognition or promotion.	1.89*
14.	I will like to opt for another job even at lesser salary	1.52
15.	I have control over my feelings to avoid job stress.	1.72

* Selected statements

ANNEXURE-III (B)

SELECTED ATTITUDE STATEMENTS FOR RURAL FAMILIES TOWARDS FRUIT AND VEGETABLE PROCESSING

Sr. No.	Statements	't' values
1.	I consider my job as a noble job	0.98
1.	One should start fruit and vegetable processing as there is vast scope of economic benefit.	2.28*
2.	Education and training can easily solve the problem of less knowledge regarding fruit and vegetable processing.	1.82*
3.	Starting fruit and vegetable processing is time consuming, therefore one should not waste time on it.	1.78*
4.	Starting food processing is money consuming, therefore one should not waste money on it.	2.19*
5.	How can one well ensure economic benefit by starting it.	1.79*
6.	Starting fruit and vegetable processing at commercial level is feasible in city area therefore, rural people should not start it.	2.25*

Sr. No.	Statements	't' values
7.	The knowhow of food processing is beyond the comprehension of illiterate and unskilled people, therefore one should not start it.	1.94*
8.	As the cost of raw material is high so it is not feasible for rural people.	1.06
9.	No one should agree with economic benefits by starting it.	1.37
10.	There is no harm in starting food processing because it promotes better living.	1.82*
11.	If one has enough knowledge of food processing it is easy to start even without training.	1.95*
12.	I try to accept changes and methods in food processing for better performance.	2.00*

* Selected statements

ANNEXURE-IV

HARYANA AGRICULTURAL UNIVERSITY

Dr.(Mrs.) U.Varma,
Assoc. Prof. and Head,
Deptt. of Extension Education
College of Home Science

Dear Sir/Madam,

    Ms. Suman Bhatti, a Ph.D. student in this department has taken up a study entitled, "Role analysis and linkages of fruit and vegetable processing organisations and institutions by rural families" under my guidance. An inventory for measurement of knowledge of rural women about fruit and vegetable processing is to be developed for the study. A list of items (enclosed herewith) has been prepared by consulting pertinent literature and experts on the subject.

    Considering your long and rich experience in this area, I am approaching you to act a judge. You are requested to judge each of the item of the attached inventory on a 3 point continum for its suitability of the scale.

    As a judge I expect you to examine each item regarding the fruit and vegetable processing for its inclusion in the scale. In case you find that inventory is not exhaustive please feel free to add any number of items to it. Also kindly rate those additional items on 3 point continum of suitability.

I will be highly ovliged and grateful for your kind help in this research project for the most valuable judgement .

Wigh regards,

Sincerely yours,

Sd/-

(U. Varma)

ANNEXURE-IV (A)

KNOWLEDGE STATEMENTS REGARDING FRUIT AND VEGETABLE PROCESSING

Sr.	Statements	Sutability		
		Most Suitable (3)	Suitable (2)	Least Suitable (1)
1.	Why there is a need of preser-vation?			
	- to enjoy the taste of fruits and vegetables in off season.			
	- To avoid wastage during peak period when it is in abundance.			
	- To avoid spoilage			
	- Any other			
2.	What are the various methods of preservation ?			
	- Drying			
	- Freezing			
	- Fermentation			
	- Use of salt			
	- Use of sugar			
	- Use of chemical preser-vatives			

Sr.	Statements	Sutability		
		Most Suitable (3)	Suitable (2)	Least Suitable (1)

3. Can you name the various products prepared by preserving fruits and vegetable?
   - Pickles
   - Jam
   - Jelly
   - Squashes
   - Syrups
   - Murabba
   - Ketchup
   - Dried fruits and vegetables

4. What are the important points to be considered while preserving fruits/vegetables?
   - Sorting of fruits/vegetables
   - Washing
   - Avoiding contamination
   - Maintain cleanliness
   - Sterlization of bottles
   - Proper packing of bottles
   - Any other

PICKLES

5. Do you prepare pickles?

6. Which ingredients in pickles help in their preservation ?
   - Oil
   - Vinegar
   - Salt

Sr.	Statements	Sutability		
		Most Suitable (3)	Suitable (2)	Least Suitable (1)
7.	What type of care should be done after pickling ?			
8.	What are the methods to prevent cloudy fluids and softness in pickles ? - Soak the vegetables in brine for sufficient time. - Use sufficient brine solution. - Cover the pickle properly with a layer of oil or brine.			

JAMS /JELLY

Sr.	Statements			
9.	Do you prepare jam /jelly ?			
10.	What is the temperature of jam at the end point ? - 100°C - 105°C - 110°C			
11.	What is pectin ?			
12.	What are commercial sources of pectin ?			
13.	What are the causes of a poor setting in jellies ?			

SYRUPS /SQUASHES

Sr.	Statements			
14.	Have you ever prepared syrup /squash ?			
15.	Why is a head space of about 3-4 cm. left at the top of a bottle ?			

Sr.	Statements	Sutability		
		Most Suitable (3)	Suitable (2)	Least Suitable (1)

16. Which ingredients acts as a main preservative in syrup and squashes ?

17. What are the functions of citric acid in syrup and squashes ?

18. What are the major defects/ spoilage in syrups/squashes?
   (a) Sediment in bottle after storage
   (b) Off or mushy flavour
   (c) Mould on cork or on top of syrup.

19. How can we prevent the spoil-age in syrups/ squashes ?
   (a) Filler the juice properly.
   (b) Use good quality fruits
   (c) Use good quality corks.
   (d) Sterillize the cork before use

PRESERVES (MURABBA)

20. What is a preserve ?

21. Do you know how to prepare a murabba ?

22. What is the difference between a jam and murabba ?

23. Why is fruit boiled in two per cent alum solution ?

Sr.	Statements	Sutability		
		Most Suitable (3)	Suitable (2)	Least Suitable (1)

24. What is the use of citric acid in a preserve ?

25. What are major causes of spoilage in preserves, jams and jellies ?
   (a) Of over ripe fruit
   (b) Under cooking or over coo
   (c) Warm or damp storage
   (d) Not enough sugar used.
   (e) Failure to remove scum.

KETCHUP

26. Are you preparing ketchups at your home ?

27. How will you determine the finishing point in ketchup ?

DRYING

28. Do you dry fruits and vegetables ?

29. What is the general method of drying vegetables ?

30. What precaution we should take while drying vegetable?
   (a) Vegetables should be tender, firm and of good quality.
   (b) Of overripe and too young vegetables should not be dried.

Sr.	Statements	Sutability		
		Most Suitable (3)	Suitable (2)	Least Suitable (1)
	(c) Should be dried rapildly after harvesting.			
31.	Do you know the other methods of drying ?			
	(i) Indirect drying (Solar dryer)			
32.	Have you ever heard about solar dryer ?			
33.	Have you ever seen solar dryer ?			
34.	Why it is called solar dryer ?			
	- Because fruits and vegetables dried with the help of indirect sun light.			
35.	Is solar dryer easy to operate?			
36.	How solar dryer looks like ?			
	- It looks like a box.			
37.	Do you know of what material it is made of ?			
38.	How many trays can be kept at a time ?			
	- Four			
39.	Where it should be kept ?			
	- Where Maximum sun light falls.			
40.	What are the merits of solar dryer ?			
	- Free from dust.			
	- Free from insect infestation			

Sr.	Statements	Sutability		
		Most Suitable (3)	**Suitable** (2)	**Least Suitable** (1)
41.	Do you know about different institutions giving credit facilities ?			
42.	Do you know how to apply for loans ?			
43.	Do you know about institutions giving guidance and training ?			

ANNEXURE-IV(B)

KNOWLEDGE TEST-FRUIT AND VEGETABLE PROCESSING

List of original statements with their difficulty indices, discrimination indices and point biserial correlation.

Sr. No.	Statement	Diffi-culty Index	Discri-mina tion Index	Point bserial corre-lation lation
1	2	3	4	5
1.	Why there is a need of preservation ? - To enjoy the taste of fruits and vegtables in of season - To avoid wastage druing peak period when it is in abundance - To avoid spoilage - Any other	73.0*	0.46	0.62*
2.	What are the various methods of preservation ? - Drying - Freezing - Fermentation - Use of salt - Use of sugar - Use of chemical preser-vatives	69.0*	0.46*	0.55*

1	2	3	4	5
3.	Can you name the various products prepared by preserving fruits and vegetables ? - Pickles - Jams - Jellies - Syrups - Squashes - Murabba - Ketchup Dried vegetables and fruits	25.0*	0.32*	0.50*
4.	What are the important points to be considered while preserving fruits/vegetables - Sorting of fruits and vegetables - Washing - Avoiding contamination - Maintain clean liness - Sterilization of bottles - Proper packing of bottles - Any other	31.0*	0.72*	0.58*

PICKLES

5.	Do you prepare Pickles ?	72.0*	0.46*	0.49*
6.	Which ingredients in pickles help in preservation ? - Oil - Vinegar - Salt	25.0*	0.43*	0.71*
7.	What type of care should be done after pickling ?	59.0*	0.64*	0.35*

1	2	3	4	5
8.	What are the methods to prevents cloudy fluid and softness in pickles ? - Soak the vegetables in brine for sufficient time. - Use sufficient brine selution - Cover the pickles properly with a layer of oil or brine Jam/Jelly	25.0*	0.39*	0.49)
9.	Do you prepare jam/jelly ?	85.0	0.28	0.23
10.	What is the temperature of jam at the end point ? - 100°C - 105°C - 110°C	15.0	0.28	0.25
11.	What is pectin ?	12.0	0.90	0.22
12.	What are the commercial sources of pectin ?	80.0	0.26	0.15
13.	What are the causes of a poor setting in Jellies ?	39.0*	0.43*	0.39*
14.	Have you ever prepared syrups/squashes.	25.0*	0.58*	0.49*
15.	Why is a head space of about 3-4 cm. left at the top of a bottle.	60.0*	0.46*	0.38*
16.	Which ingredient acts as a main preservative in syrup and squashes?	60.0*	0.46*	0.38*
17.	What are the functions of citric acid in syrup and squashes ?	25.0*	0.46*	0.60*

1	2	3	4	5
18.	What are the major defect/ spoilage in syrup/squash. - Sediment in bottle after spoilage - Mould on cork or on top of syrup.	28.0*	0.43*	0.39*
19.	How can we prevent the spoilage in syrups/squashes ? - Filter the juice properly - Use good quality fruits - Use good quality corks. - Sterilize the cork before use.	35.0*	0.48*	0.52*

PRESERVES / MURABBA

1	2	3	4	5
20.	What is a preserve ?	16.0*	0.23	0.25
21.	Do you know how to prepare a murabba ?	70.0*	0.58*	0.49*
22.	What is the difference between a jam and murabba.	12.0	0.26	0.15
23.	Why is fruit boiled in two per cent alum solution ?	15.0	0.23	0.25
24.	What is the use of citric acid in a preserve ?	35.0*	0.38*	0.47*
25.	What are the major causes of spoilage in p[reserves, jams and jellies ? - Overripe fruits - Under cooking or over cooking - Warm or damp storage - Not enough sugar used - Failure to remove slum	71.0*	0.65*	0.76*

1	2	3	4	5
	KETCHUP			
26.	Are you preparing ketchups at your home ?	35.0*	0.58*	0.49*
27.	How will you determine the finishing point in ketchup ?	15.0	0.28	0.23
	DRYING			
28.	Do you dry fruits and vegetables?	25.0*	0.52*	0.49*
29.	What is the general method of drying vegetables ?	15.0*	0.28*	0.28*
30.	What precautions we should take while drying fruits and vegetables ? - Fruits and vegetables should be tender, firm and of good quality - Should be dried rapidly after harvesting	44.0*	0.65*	0.49*
31.	Do you know the other methods of drying ? - Indirect drying (Solar dryers)	35.0*	0.48*	0.52*
32.	Have you ever heard about solar dryer ?	44.0*	0.52*	0.43*
33.	Have you even seen solar dryer ?	15.0	0.28	0.15
34.	Why it is called solar dryer ? - Because products are dried in this with the help of sun light	71.0*	0.65*	0.76*
35.	Is solare dryer easy to perate ?	12.0	0.28	0.23
36.	How solar dryer looks like ?	16.0	0.23	0.25

1	2	3	4	5
37.	Do you know of what material it is make of ?	12.0	0.26	0.15
38.	How mahy trays can be kept at a time ?	60.0*	0.48.0*	0.52*
39.	Where it should be kept?	59.0*	0.63*	0.56*
40.	What are the merits of solar dryer ?	43.0*	0.48*	0.52*
41.	Do you know about different institutions giving credit facilities	79.0*	0.65*	0.76*
42.	Do you know how to apply for loans ?	40.0*	0.38*	0.47*
43.	Do you know about institutions giving guidance and training ?	28.0*	0.43*	0.49*

* Statements selected for knowledge lest.

ANNEXURE-V

ROLE ANALYSIS AND LINKAGES OF FRUIT AND VEGETABLE PROCESSING ORGANISATIONS AND INSTITUTIONS BY RURAL FAMILIES

(Interview shedule for organisation)

General Information

1. Name of the organisation

2. Type of organisation

3. Total area :
   Building (Plinth area) :
   Workship (Production area) :

4. Working hours (timings) :

5. Approximatesale per year :

6. Approximate sale per year :

7. What are the inputs in your organisation

	Funding agency		
	Govt.	**Private**	**Cooperative**

A (a) Capital
       (i) Initial capital
       (ii) Running capital
   (b) How the capital was raised ?
   Govt.Loan
   Govt. Subsidy
   Cooperative
   Pesonal

B. Machinery and Equipment
   Total number of machines and equipment :
   Costing more than Rs. 5000/-
   Costing more than Rs. 10,000/-
   Costing more than Rs. 20,000/-
   Costing more than Rs. 50,000/-
   Costing more than Rs. 1 lakh

8. What are the major aims of your organsation ?

9. What are the major roles performed by your organisation
   Production
   Training
   Education
   Research and Development
   Production and training
   Any other

10. What are the various status in your organisation ?

Status	No. of persons		Educational
	Male	Female	Qualification
1.			
2.			
3.			
4.			

11. What are the roles of each status ?

12. What constraints do you feel in performing your roles ?

13. What are the major fruit and vegetable products being processed in your organisation ?
    Jams
    Jellies
    Juices
    Syrups
    Pickles
    Any other (specify)

14. What are the major processing methods being used?
    1. Use of sugar
    2. Use of salt

3. Use of chemical preservatives.
4. Use of other preservatives.
5. Canning
6. Drying
7. Freezing
8. Fermentation
9. Bottling

15. What are the outputs in your organisation ?

Processed products	Total production (Quantity)	Total cost

16. How many persons obtained training from your organisation?

Total no. of persons		Type of training	Period of training
Male	Female		

17. What are the major processing activities performed by male and female ?

	Total man power	Male	Female
Selection			
Washing			
Grading			
Peeling			
Chopping			
Blanching			
Preserving			
Packaging			
Storage			

18. From where do you purchase raw material ?
Local market
Nearby city
Far off city
Outside district

Outside state

Outside Country.

19. What constraints do you face while purchasing raw material?

20. From where do you purchase machines and equipments ?

Local market

Nearby city

Far off city

Outside district

Outside district

Outside country

21. What constraints do you face while purchasing machines and equipments.

22. Education :

Illiterate	-	0
Primary	-	1
Middle	-	2
Matric	-	3
Technical/vocational edu.	-	4
Graduate	-	5
Post graduate	-	6

23. Service Experience

Total service experience at the present job	Yrs.	Months

24. Attitude toward job.

(Given below are some statements regarding job in fruit and vegetable processing organisation. Each statement has five possible responses for "Strongly agree" to "Strongly disagree". You are requested to check ( ) the appropriate response category against each statement expressing your agreement or disagreement with the statement. Nothing is right or wrong, we are just interested in your feelings. Your reply would be kept strictly confidential. Please give frank opinion).

Statements	SA*	A	UD	D	SD

1. To me nothing is more satisfying than working in fruit and vegetable processing organisation.

2. I always try to accept changes for better performance.

3. The sooner I get out of the job in fruit and vegetable processing industry the better it is.

4. Working for fruit and vegetable processing is very important for national development.

5. Working for fruit and vegetable processing is the need of time.

6. Working in fruit and vegetable processing is not rewarding.

7. Working in fruit and vegetable processing is very satisfying for me.

8. There is no use of working in fruit and vegetable processing organisation as there is no chance of making money.

9. I consider my job challenging in the era of technological advancement.

10. I try to put my head and heart in to my job without bothering about recognition or promotion.

---

25. Job Satisfaction

Proscribed below are some statements on different aspects of job life againts each statement on agreement-disagreement continuum with categories,strongly agree (SA), agree (A), undecided (UD), disagree (D) and strongly disagree (SD) are given. You are requested to check ( ) anyone categories against each statement, expressing the extent of your satisfaction - dissatisfaction with that aspect.

Statements	SA	A	UD	D	SD
1. I feel that I should get more pay according to my qualifications.					
2. I am sure to get promotion in due time without any problem.					
3. Due incentives and increments are not given in my organisation for good work.					
4. Usually, I am dissatisfied as I don't get enough facilities to carry out my job.					
5. I get work according to my capabilities and ablities.					
6. Usually, I get bored in my job.					
7. There is ample scope in my job to prove my excellence in work.					
8. My organisation provides equal and ample opportunities for participating in training, seminars etc.					
9. In my organisation, all jobs are temporary all the time I am afraid that I will be turned out any time.					
10. I get recognition for my good work.					
11. My job gives me no status					
12. I feel that my job is not according to my taste.					
13. Here, I get good medical facilities for my family and myself.					
14. I enjoy my work more than even my leisure time.					
15. My job is very tiring					
16. I feel that seniors always evaluate the work on the merit and performance not on anything else.					

(Contd.)

Statements	SA	A	UD	D	SD
17. My superiors listen to my problems with sympathy.					
18. Here we always work together in a team.					
19. Here I never get the guidance from my superiors whenever I need it.					
20. I am always afraid that I way be transferred any time.					

26. Organisational climate

Statements	SA	A	UD	D	SD

1. **Orientation**

   (a) People here are more concerned about following rules and procedures.

   (b) The highest concern in the organisation is about the development of people.

   (c) Consolidating one's own personal position and influence seems to be the main concern.

   (d) The dominant concern here is to maintain friendly relations with others.

   (f) The main concern in the organisation is to develop competence and expert functions.

2. **Inter-personal Relationship**

   (a) Here competent persons are respected.

   (b) The atmosphere here is very friendly and people spend enough time in informal social relations.

(Contd.)

Statements	SA	A	UD	D	SD

(c) There are strong cliques in the organisation.

(d) Business like relationship prevail herepeople are warm, but get together mostly for work.

(e) People have strong associations with their supervisors and mostly look for suggestion and guidance from them.

(f) People have great concern for one another, and help others spontaneously when such help is needed.

3. Supervision

(a) Supervision here is usually to check mistakes and 'catch' the person.

(b) The main spirit of supervision is to guide the person supervised.

(c) Supervisors are genuinely concerned about the growthand development of their subordi-nates.

(d) Supervision is effective in creating a climate of commitment and achievement.

(e) Supervisors try to use their expertise and competence rather than their positions in influencing their subordinates.

(f) Supervisors are very friendly and find it difficult to communicate strong feedback to their subordinates.

(Contd.)

Statements	SA	A	UD	D	SD

4. Communication

(a) Communication is usually one way from topn to down in the organisation.

(b) Only those who know each other well get the information first.

(c) People ask for informationfrom those who seem to know the subject well.

(d) Relevant informationis available to all those who need and can use such information.

(e) People taking initiative in communicating show concern for others.

(f) Communication is often selective people usually hold back some crucial information as a way to control.

5. Decision Making

(a) In decision making people usually involve their close friends.

(b) Decisions are usually made at the top and communicated downward.

(c) Decisions are made at appropriate level by those who are concerned and those who have relevant information and resources.

(d) In decision making only a few chosen participate.

(Contd.)

Statements	SA	A	UD	D	SD

(e) Decisions are made and influenced by competent and knowledgeable persons.

(f) Decisions are made by the concerned persons but others support and provide necessary help.

6. Trust

(a) Only a few persons are trusted and they are quiet influential.

(b) Only close friends are trusted

(c) Only the superiors (or subordinates) are trusted more.

(d) Generally people can rely on help and support of others.

(e) Generally people have trust in the abilities of others.

7. Managing problems

(a) People solve problems individually

(b) Experts are consulted, they play an important role in solving problems.

(c) People usually refer the problems and look for solutions to their seniors.

(f) People with higher sense of responsibility mainly take initiative in solving problems.

8. Managing Mistakes

(a) The person making a mistake is not rejected he is shown warmth and his friends support him.

(Contd.)

Statements	SA	A	UD	D	SD

(b) People try to defend themselves when they make mistakes.

(c) The person is given help in analysing the mistake to prevent it in future.

(d) A mistake is seen as an experience from which lessons are learnt to prevent failure in future.

(e) Superiors keep on guiding their subordinates to prevent mistakes or correct mistakes when they occur.

(f) Help of experts is sought in analysing and preventing mistakes.

9. **Management conflicts**

(a) People focus attention on effectiveness of their roles, and being less concerned with others roles, conflicts do not arise.

(b) Conflicts are usually avoided and people refer to friendly atmosphere to help avoid conflicts.

(c) Arbitration or third party intervention (usually by experienced persons or seniors) mis sought and used.

(d) Those who are stronger, force their point of view.

(e) Concern is shown and people try to help each other in resolving conflicts.

(f) The problem is analysed and solution worked out in a dispassiopnate way.

(Contd.)

Statements	SA	A	UD	D	SD

10. Managing rewards

(a) Only efficiency in work is awarded.

(b) General competence and knowledge is recognised more.

(c) Loyalty is rewarded more than any thing else.

(d) The organisation rewards those who are able to get things done by others.

(e) Personal relations play a great role in the reward system.

(f) The organisation recognises those who are able to get things done by others.

11. Risk taking

(a) Very little risk is taken by people here.

(b) People here take very high risk.

(c) Risk taking is fairly high, but persons who seem to know the subject are involved in decisions.

(d) People take risk with the approval of their superiors.

(e) People take risk with confidence about the help and support they will get from their colleagues.

(f) Calculated risk is taken by individuals.

(Contd.)

27. Professional commitment

Below are given some statements. Each statement has five possible responses viz. Very least important upto very greatly important. Kindly check ( ) aginst each statement to a response to which you agree.

Statements	VLI	LI	I	GI	VGI
1. To make use of my knowledge and skill in my job.					
2. To increase my knowledge in the particular field.					
3. To work with colleagues of high technical competenc.					
4. To build my professional reputation.					
5. To work on diffcult aqnd challenging problems.					
6. To contribute new ideas to my field.					

28. Organisational Commitment

Statements	SA	A	UD	D	SD
1. I am willing to put in a great deal of efforts beyond what normally is expected in order to help this organisation be successful.					
2. I talk up this organisation to my friends as a great organisation in to work for					
3. I feel very little loyality to this organisation.					
4. I would accept almost any type of job assignment in order to keep working for this organisation.					
5. I find that my values and organisational values are similar.					
6. I am proud to tell others that I am part of this organisation.					

(Contd.)

Statements	SA	A	UD	D	SD
7. I would just as well be working for a different organisation as long as the types of work were similar.					
8. This organisation really inspires the best in me in the way of job performance.					
9. It would take vgery little change in my present circumstances to cause me to leave this organisation.					
10. I am extremely glad I choose this organisationto work for other I considering at the time I joined.					
11. There is not much to be gained by sticking with this organisation indefinately.					
12. Often, I find it difficult to agree with this organisation's policies on important matters relating to its employees.					
13. I really care about the fate of this organisation.					
14. For me, this is the best of all organisations for which to work.					
15. Deciding to work for this organisation was a definte mistake on my part.					

29. Inter-organisation linkages

It is considered that fruit and vegetable processing organisations and institutions are communicating aqnd coord nating with technical and financial institutions. Therefore, let it be known that what are the agencies and institutions you contacted and mode of communication used. Please check () the agency, institution, personnel and mode you actually used.

CONSTRAINTS

What constraits do you face while working ?

Administrative constraints   **Yes/No**

1. Centralised planning
2. Scheme not effective at implementation stage
3. Delegationof responsibility without power
4. More of formalities and paper work
5. Lack of feed back and monitoring.
6. Lack of infrastructural facilities.

Staffing

1. Staff indequacy
2. Staff incompetency
3. Lack of refresher courses
4. Limitationof lower staff.

Communicational

1. Lack of communicational facilities
2. Lack of mass media communication
3. Rare possibility of personal contacts.

Linkages and coordination

1. Lack of coordinated approach
2. Clash in objectives/targets of different organisations
3. Non-recognitionbyh other organisations.
4. Political interference.
5. No control on personnel of other departments

Budgeting

1. Lack of funds
2. Poor allocationof budget for different purposes.
3. Limited travel grants.

Raw Material
Yes/No

1. Lack of supply of raw material for unintrupted production

2. Procurement of raw material is time consuming and require diligent selection.

3. Variations in raw material prices at different places.

4. Lack of sufficient stock of raw material inperiod of short supply and anticipated price change.

Marketing

1. Competition from established and larger unit in the production line.

2. Delayed disposal of produce.

3. Difficulty in getting money from buyer after sale.

4. Lack of sufficient finished goods for smooth sales operation and efficient customer service.

ANNEXURE VI

ROLE ANALYSIS AND LINKAGES OF FRUIT AND
VEGETABLE PROCESSING ORGANISATIONS AND
INSTITUTIONS BY RURAL FAMILIES.

**(Interview schedule for institutions)
General Information**

1. Name of the respondent

2. Father's/Husband's Name socio-personal variables

3. Age Actual age
   - (i) Below 25 yrs.   1
   - (ii) 25-50 yrs.   2
   - (iii) Above 50 yrs   3

4. **Caste**

   Low
   - (i) Chamar, Bhangi, Dom   1
   - (ii) Jhimar, Khati, Dhobi, badi   2

   Middle
   - (iii) Lohar, Kumhar, Darji, Nai   3
   - (iv) bania, Sonar, Ahir, Jolaha   4
     Saini, Arora
   - (v) Brahmin   5

   High
   - (vi) Jat, Rajput, Bishnoi   6

5. Family Education status :

Illiterate	0
Primary	1
Middle	2
High School	3
Technical/vocational educatio	4
Graduate	5
Post-graduate	6

Sr. No.	Relation with respondent	Educational Qualification	Score
	Self		
	Father		

6. Type of family

(i) Nuclear	1
(ii) Joint	2

7. Size of family

(i) Small (upto 5 members)	1
(ii) Medium (upto 10 members)	2
(iii) Large (above 10 members)	3

8. Status in family

For rural women

Daughter	1
Daughter-in-law	2
Mother-in-law	3

For rural men

Unmarried son	1
Married son	2
Father-in-law	3

9. social participation

(i)	Nil	0
(ii)	Member one organisation	1
(iii)	Member of more than one organisation	2
(iv)	Office holder	3
(v)	Wider public leader	4

10. Urban contact

How often do you visit towan or citis for social personal or any other purpose.

Frequency of visit	Score
Never	0
Sometimes (between 6 months)	1
Frequently (six months & above)	2

11. Extension contact

How often during the last one year did you come in contact with the following personnel for getting information pertaining to fruit and vegetable processing.

Extension personnel	Frequency of contact		
	5-6 contacts Mostly	3-4 Contacts Sometimes	1-2 Contacts Never

12. Type of unit

Commercial scale	2
Home Scale	1

13. Total area of building

14. Approximate initial capital

15. Approximate runing capital

16. Approximate sale per year

17. Did you obtainfinancial assistance   **Yes/No**
   from other sources ?

   If Yes.

   (a) Percentage of   25%   50%   75%
       assistance obtained

   (b) Mentionthe source and their approx share in terms of %.
       Score   %
       Family members
       Relatives &frientds Banks
       Private money lender
       NGO's
       Govt, agency

18. Machinery and equipment possession

 1. Balance   l
 2. Cutting knife   l
 3. Peelter-cum-cover   l

Juice extraction

 4. Hand juice machine   l
 5. Electric juic emachine   l
 6. Lime juice squeeser   l

Sieving

 7. Sieves

Heating equipments

 8. Stove   l
 9. Cooking gas burner   l

Instruments

 10. Brix hydrumeter   l
 11. Jelmeter   l
 12. Thermometers   l

Canning

 13. Home can sealer   l
 14. Can-cum-cork opener   l
 15. Crown corking machine   l
 16. Crown corks   l
 17. Pressure cooker (coner)   l

Containers

 18. Bottles

19.	Cans	1
20.	Plastic jars	1

Enameled vessels

21.	Basins	1
22.	Caps	1
23.	Funnel	1

Other accessories

24.	Glass tumblers	1
25.	Measuring cylinder	1
26.	Beakers	1
27.	Cooking pan	1
28.	Bottle brushes	1
29.	Rubber gloves	1
30.	Wooden spoons	1
31.	Pressure cooker	1
32.	Drying trays	1

Chemicals

33.	Citric acid	1
34.	Edible colours	1
35.	Essence	1
36.	Potassium metabisulphite	1
37.	Sodium bemoate	1
38.	Glacial acetic acid	1
39.	methylated spirit	1

Miscellaneous

40.	Dusters	1
41.	Muslin cloth	1
42.	Vim.	1
43.	Plastic scrubber	1

19. Availability of technical/managerial guidance.

No. guidance	0
Technical guidance	1
Managerial guidance	2

20. Availability of Machinery and equipment

Not neede	0
Available at selected place	1
Easily available	2

21. Raw material

Not available in time	0
Available at few places	1
Easily available	2

22. Attitude towards fruits and vegetable processing

Statements	SA	A	UD	D	SD
1. One should start fruit and vegetable processing as there is vast scope of economic benefit.					
2. Education and training can easily solve the problem of tless knowledge regarding fruit and vegetable processing.					
3. Starting fruit and vegetable processing is time consuming, therefore, mone should not waste time on it.					
4. Starting food processing is money consuming, therefore, one should not waste money on it.					
5. How can one well ensure economic benefit by starting it.					
6. Startin gfruit and vegetable processing at commercial level is feasible incity area, therefore, rural people should not start it.					
7. The knowledge of food processing is beyond the comprehension of illiterate and unskilled people, there fore one should not start is.					
8. There is no harm in starting food processin gbecause it promotes better living.					
9. If one has enough knowledge of food processin it is easy to start even without training.					
10. I try to accept changes and methods in food processin gfor better performance.					

Economic Motivation

Sr. No.	Statement	Most Like 2	Least Like 1
1.	All I want from my processing work is to make just a reasonable living for the family.		
2.	In addition to making reasonable amount of profit, the enjoyment in fruit and vegetable processin gis also important for me.		
3.	I would invest in fruit and vegetable processin gproduction to the maximum to gain large profits.		
4.	I would not hesitate to borrow anyamount of money in order to manage my fruit and vegetable processin gwork properly.		
5.	Instead of adopting improved technologies which cost more I follow the routine fruit and vegetable processing practices.		
6.	It is not only monetary profit but the enjoyment of work which gives me satisfaction for proper working of fruit and vegetable processing.		
8.	My main aim is maximising monetary profit from fruit and vegetable processing through processed food which is simply consumed by my family.		
9.	I avoid excessive borrowing of money for fruit and vegetable processing.		

Risk Orientation

Statements	SA	A	UD	D	SD
1. Person should rather take more of a chance in making a big profit than to be content with a smalle rbut less risky profit.					
2. Those people who are willing to more risk in adopting new innovation than the average people are always better off.					
3. It is good for one to take risk in adopting new innovations when one knows that it will be for his/her benefit.					
4. Trying an entirely new innovations by rural women involves risks, but is worth of it.					
5. One should not adopt new innovations oneself unless other had adopted it.					
6. Adopting new innovations involves risk but they are worth it.					

Constraints encountered byrural families   Yes/No

General personnel constraints

1. Excessive burden of work and responsibility
2. Health problem
3. Lack of leisure time and other activities
4. Lack of systematic planning and working
5. Lack of competencfe in handling activities.
    (a) Managerial
    (b) Financial
    (c) Sale
    (d) Production

6. Lack of information about :
   (a) Agencies and institutions working for fruit aqnd vegetable processing.
   (b) Various schemes run by Govt.
   (c) Raw material availability
   (d) Machinery and equipment aqvailability
   (e) Marketing
   (f) Various improved technologies
   (g) Loaning schemes and procedures of financial institutions.

Socio-psychological constraints

1. Lack of interest
2. Non cooperativ attitude of family members.
3. Lack of recognitionand appreciationin the family.

Financial constraints

1. Lack of sufficient working capital
2. Frequent and higher need of finances.

Technological constraints

1. Non availability of modern technologies
2. Availability of technologies /machinery & equipment at a higher price or at a distant place.
3. Any other (specify)

Raw material constraints

1. Lack of supply of raw material for unintrupted production.
2. Procurement of raw material is time consuming and requires diligent sel ction.
3. Variations in raw material prices at different places.

Marketing constraints

1. Lack of marketing intelligence
2. competition from established and larger units in the productionline.
3. Delayed disposal of produce.
4. Difficulty in getting money from buyer after sale.

Role performance of rural families in fruit and vegetable processing

How often do you perform following fruit and vegetable processin gtasks ?

Activities	Extent of performance		
	Regularly	**Some times**	**Never**

Managerial roles

1. Making a plan of work
2. Establishing processing priorities.
3. Identifying need for inputs such as raw material.
4. Arranging for finances.
5. Making contacts for finances.
6. Maintaining records for production, market price and sale.
7. Seeking information about loaning schemes.
8. Securing informationfrom experi-mental stations.
9. Modifying plans according to season.
10. Purchasing raw materials.

Processing roles

1. Selecting the fruits and vegetables.
2. Washing fruits and vegetales.
3. Chopping for processing.
4. Blanching fruits and vegetables
5. Preserving
6. Packaging the products.

Activities	Extent of performance		
	Regularly	**Some times**	**Never**

Marketing roles

1. Identifying segments/targets
2. Analysing competetion
3. Planning according to season.
4. Packaging according to market
5. Checking the price
6. Deciding for promotion
7. Deciding for distribution.

Linkages

It is considered that rural families engaged in fruit and vegetable processin gare communicating and cordrinating with technical and financial institutions, rural families of their own and other villages and with traders. Therefore let it be known that what are the agencies and institutions you contacted and mode of communication used and what was the extent of use of different methods. Please check ( ) the agency, institutioon, personnel, media/method adopted alongwith its extent of use during the last one year.

	EXTENT OF COMMUNICATION						
Institution	Scientists of the SAUs	Training unit personnel	Banks	Families of own village	Families of other village	Traders	Cooperatives
	M,S,N,	M,S,N,	M,S,N,	M,S,N,	M,S,N,	M,S,N,	
Rural Families							

MODE USED BY FRUIT AND VEGETABLE PROCESSING INSTITUTION FOR LINKAGES

Organisation Instituitions	Meetings	Panchayat meetings	Letter	Personal visit	Trainings	Shows	Any other
	M,S,N,	M,S,N,	M,S,N,	M,S,N,	M,S,N,	M,S,N,	M,S,N,
Scientists of the SAUs.							
Training unit personnel							
Banks							
Families of Own village							
Families of other villages							
Traders							
Cooperatives.							

FRUIT AND VEGETABLE PROCESSING ORGANISATIONS AND INSTITUTIONS

By

SUMAN BHATTI

Major Advisor : **Dr. (Mrs.) U. Varma.**
Assoc. Profesor,
Department of Home Science
Extension Education,
CCS Haryana Agricultural University,
Hisar -125004, Haryana

(An abstract of thesis submitted in partial fulfilment of the requirements for the degree of Doctor of Philosophy, CCs haryana Agriculatural University, Hisar).

The present study was conducted to analyse the role performance and linkages of fruit and vegetable processing organisations and institutions. The study was conducted in three districts, viz.. Hisar, Panipat and Faridabad. The requisite information was collected with the help of an interview schedule from eight organisations and 65 rural families working in these organisations and also engaged in fruit and vegetable processing activities either at home scale or at commercial level.

The findings revealed that all the organisations were performing the role of productiononly. Only one organisation was performing the role of training, research and development and production. Number of women employee was less as compared to men. They were employed only for manual works. Inter-organisational linkages were negligible. All the organisations were working in isolation.

At the institutional level most of the respondents were working at commercial level. All the families were practising drying and use of salt as in managerial and marketing role was poor, whereas average performance was observed in processing activities. The significant difference was observed in the performance of both men and women in fruit and vegetable processing activities. Lack of appropriate technology, lack of funds, dual responsibilities, non-availability of raw material and getting sale-proceeds from buyer belatedly after sale were some of the major constraints faced by rural families.

On the basis of judges consensus, evaporative cooling system, drying and tray packing emrged as the appropriate technologies for post-harvest, processing and packaging of fruits and vegetables. And based on secondary data, zero-energy cool chamber and drying were the appropriate technologies for fruits and vegetables. A comparative study on natural convection solar dryer (chimney) and open sun drying revealed the quality of the dried product to be good.

The data further revealed the eduacation, caste, urban contact, type of unit, availability of raw material, training, machinery and equipement possession and economic motivationhad significant positive correlationwith role performance and linkages of rural families. Individually, type of unit exhibited significant regression co-efficients towards role performance and linkages of rural families.

AUTHOR INDEX

SUBJECT INDEX